T0292048

THE GOOD DRONE

ACTING WITH TECHNOLOGY

Bonnie Nardi, Victor Kaptelinin, and Kirsten Foot, editors

Tracing Genres through Organizations: A Sociocultural Approach to Information Design,
Clay Spinuzzi, 2003

Activity-Centered Design: An Ecological Approach to Designing Smart Tools and Usable Systems, Geri Gay and Helene Hembrooke, 2004

The Semiotic Engineering of Human-Computer Interaction, Clarisse Sieckenius de Souza, 2005

Group Cognition: Computer Support for Building Collaborative Knowledge, Gerry Stahl, 2006

Acting with Technology: Activity Theory and Interaction Design, Victor Kaptelinin and Bonnie A. Nardi, 2006

Web Campaigning, Kirsten A. Foot and Steven M. Schneider, 2006

Scientific Collaboration on the Internet, Gary M. Olson, Ann Zimmerman, and Nathan Bos, editors, 2008

Acting with Technology: Activity Theory and Interaction Design, Victor Kaptelinin and Bonnie A. Nardi, 2009

Digitally Enabled Social Change: Online and Offline Activism in the Age of the Internet, Jennifer Earl and Katrina Kimport, 2011

Invisible Users: Youth in the Internet Cafés of Urban Ghana, Jenna Burrell, 2012

Venture Labor: Work and the Burden of Risk in Innovative Industries, Gina Neff, 2012

Car Crashes without Cars: Lessons about Simulation Technology and Organizational Change from Automotive Design, Paul M. Leonardi, 2012

Coding Places: Software Practice in a South American City, Yuri Takhteyev, 2012

Technology Choices: Why Occupations Differ in Their Embrace of New Technology, Diane E. Bailey and Paul M. Leonardi, 2015

Shifting Practices: A Reflective Inquiry into Technology, Practice, and Innovation, Giovan Francesco Lanzara, 2016

Heteromation, and Other Stories of Computing and Capitalism, Hamid R. Ekbia and Bonnie Nardi, 2017

The Good Drone: How Social Movements Democratize Surveillance, Austin Choi-Fitzpatrick, 2020

THE GOOD DRONE

How Social Movements Democratize Surveillance

AUSTIN CHOI-FITZPATRICK

The MIT Press
Cambridge, Massachusetts
London, England

This book was set in Adobe Garamond Pro and Berthold Akzidenz Grotesk by Westchester Publishing Services.

Library of Congress Cataloging-in-Publication Data

Names: Choi-Fitzpatrick, Austin, author.
Title: The good drone : how social movements democratize surveillance / Austin Choi-Fitzpatrick.
Description: Cambridge, Massachusetts : The MIT Press, [2020] | Series: Acting with technology | Includes bibliographical references and index.
Identifiers: LCCN 2019034699 | ISBN 9780262538886 (paperback)
Subjects: LCSH: Social movements--Technological innovations. | Technology--Social aspects. | Technology--Political aspects.
Classification: LCC HM881 .C4445 2020 | DDC 303.48/3--dc23
LC record available at https://lccn.loc.gov/2019034699

To Naomi Yoder, who sleeps under the stars
To Aila Pax, who flies in her dreams
To Eden Justice, who sees in the dark

CONTENTS

ACKNOWLEDGMENTS

Books are rolling accumulations of debt—some to the living, some to those without breath.

THINGS

Since this is a book about technology, let me first thank those without breath: the technologies that made the manuscript possible. Hunches for this book emerged when I purchased a DJI Phantom from an American vendor and had it shipped to my office in Hungary. A 3D-printed gimbal, a modified camera housing, and custom landing gears were needed almost right away.

Later I bought a large latex balloon from Public Lab as well as a camera, a 3D-printed apparatus for carrying that camera, and a grip-load of string. The kite I'd always had laying around, a story I tell in the second chapter. My students have built their own drones using transmitters, receivers, motors, processors, control boards, and propellers—so many propellers. We ordered these things from Amazon and Banggood.

Additional fieldwork required buckets of paint and rattle cans for public art deployed onto public spaces. Paste, markers, paint, glitter, and poster board went into posters carried at some marches, and larger-than-life puppets were deployed in other protest events. I've also used my vocal chords and my body to signal public assent and dissent.

I wrote much of this text using the software Scrivener running on a MacBook Air. Early ideas were dictated directly into Google Mail via Google's voice-to-text feature. Interviews were captured on a Motorola smartphone running Android. I conducted background research, followed hunches, and messed about using Google Search and Scholar, Quora, Wikipedia, Twitter, and, until I nominally abandoned the platform in early November 2016, Facebook.

Books from many libraries fill my office. I borrowed rolling cart number 25 from the library and never returned it (well, I never told them that). The books I liked the most I bought and marked up extensively. All research materials were stored in Dropbox (on Amazon's servers worldwide). Additional books were consumed audibly (thanks, Audible). I'm always losing earbuds, but those are important too.

Working drafts were printed on a Sharp MX3050V printer in San Diego, a Konica Minolta bizhub C224 in Budapest, and a Brother DCP-L2540DW at home. I read the markup with the aid of my eyeglasses (1.25/-3.00/92 x .75/-2.5/80) and marked up the copy with whatever pens I had at hand—I love them, but lose them. As the ideas took shape, I hailed ride-sharing services, hopped onto airplanes, and crashed in sharing-economy housing so that I could discuss my inklings in talks that required first espresso and then thumb-drives, dongles, overhead projectors, coffee, and later wine.

In manuscript form, this book underwent an open peer review on the platform PubPub. PubPub uses hosting services provided by Amazon US-East, though data is also distributed in data centers around the world. Readers commented on the draft manuscript using their browsers (60.12 percent Chrome, 22 percent Firefox, and .241 percent Internet Explorer). Later, this book went to press. Interestingly, the publishing industry has been transformed to the extent that it is near-impossible to determine where the volume in your hands was printed.

Hot off the press, though, the physical book is whisked to warehouses and stacked on a pallet in an Ingram facility, or whisked around an Amazon floor by a Kiva robot (or human equivalent), and then dispatched by UPS, FedEx, US Postal Service, Amazon Delivery person, or whatever delivery modality is at work near you now.

How you're experiencing this book is another matter altogether. One of the "yous" "experiencing" this book is an algorithm sent to drink these words into large-scale repositories (welcome!). The rest of us "yous" are humans experiencing this book in digital (audio or visual) or analog (paper or hardcover) form.

I thank all of these entities for their help in making this book possible.

PLACES

I thank bars and cafes where I wrote and edited this book. In Budapest: Espresso Embassy, Coyote Cafe, My Little Melbourne; in San Diego: Bird Rock, James, and Panikkin; in Oxford: Society Cafe, Gail's, and Handle Bar Cafe; in Kigali: Chez Lando and Question Coffee.

Spaces inside institutions like the Massachusetts Institute of Technology, University of Pennsylvania, Harvard University, Central European University's Center for Media, Data and Society, the University of San Diego, and the International Studies Association made this possible.

Server stacks and warehouse racks are places too. All hail the cloud for storing my flights of fancy, my office for letting me stick around until extra late, my home for welcoming me back no matter where I'd been, and the comfortable seat in my 2005 Honda Odyssey, where I draft and edit documents while my kids practice soccer.

All hail my local independent—and the country's oldest family-owned and operated—bookstore, Warwick's.

To all these places, I give thanks.

PEOPLE

At Central European University, Rector John Shattuck and Dean Wolfgang Reinicke forgave me for buying a drone without permission, and for getting their students stopped by the police (well, I never told them about that part).

At the University of San Diego, Provost Gail Baker, Dean Patricia Marquez, Dean Theresa Byrd, and Dean Chell Roberts have made an open access version of this book possible. At the Kroc School of Peace Studies, where I am a member of the faculty, Dean Patricia Marquez has given me the space and support needed to chase ideas. At the Shiley School of Engineering, Dean Chell Roberts has supported a number of pilot projects, while also storing a very large canister of helium for a very long period of time (well, I never told him about that part). Generous support from the National Science Foundation (IUSE/PFE RED, no. 1519453) for "revolutionizing engineering and computer science" helped underwrite some of the research described here.

At the University of Nottingham, where I am also a member of the faculty, I've had the amazing support of Pro-Vice Chancellor Todd Landman, Rights Lab Director Zoe Trodd, Rights Lab Research Director Kevin Bales, and my stalwart department chair, Ian Shaw. The Rights Lab and School of Sociology and Social Policy are endlessly supportive and patient, putting up with my crazy antics and long absences. Doreen Boyd, in particular, has helped me think in new ways about satellites.

This book wouldn't have existed without the encouragement of Phil Howard, Gina Neff (Oxford), and Kirsten Foot (Washington). They saw something interesting in this project from the get-go, and have been unfailing in their support ever since. The opportunity to spend a stretch of time writing in peace with a visiting appointment at the Oxford Internet Institute helped me get the first cut of the final draft out of the starting gate, and an invitation to submit the book to the Acting with Technology series saw it across the finish line. More broadly, I have benefitted immeasurably from the guidance and support of Rory McVeigh, Alison Brysk, Dan Myers, and Christian Davenport. Mentors who help you get started and then stick with you, or let you circle back around, are a thing of wonder. Thanks y'all!

Many of the ideas in this book first found form in conversations with students, coauthors, and student coauthors. I am particularly grateful to Gordon Hoople, Lars Almquist, and John Holland as well as an army of post-docs and graduate students, including Beth Reddy, Tautvydas Juskauskas, AlHakam Shaar, Dana Chavarria, Elizabeth Cychosz, John Paul Dingens,

Michael Duffey, Katherine Koebel, Sirisack Siriphanh, Merlynn Yurika Tulen, Heath Watanabe, Howard Smith, Nora Beglane, Nisreen Al Sabie, Patricia Cosulich, Carolyn Ross, Claire Bergstressser, and Boby Md Sabur.

Editors do the lord's work (though some have been known to moonlight for the competition). I'm pleased to say that every editor I've ever worked with has been a force for good. At MIT Press, this includes Kirsten Foot, Bonnie A. Nardi, and Victor Kaptelininin in the *Acting with Technology* series. Their patient guidance made all the difference. Editors at *Interface, The Journal of Engineering Studies, Columbia Journal of International Affairs, Al Jazeera,* and *Slate* all deserve thanks for their support in helping me develop some of the background arguments that have made their way into these pages. An earlier version of chapter 3 appeared in the volume *Visual Imagery and Human Rights Practice*, edited by Monroe Price and Sandra Ristovska, and appears here in modified form under Palgrave's generous terms of use.

There are many people to thank at MIT. My editor, Katie Helke, gets the most credit for being patient, creative, and persistent. In the window of time between when the book went through peer review with academics (thanks y'all, you know who you are!) and it hit the open waters of public readership, it went through an open review process on the PubPub platform. This innovative effort is the brainchild of Travis Rich and is implemented by a team that includes Catherine Ahearn and Gabriel Stein. I'm grateful to the PubPub gang for helping me share these early ideas with the world. My fingers are crossed this will be a living book as well.

A thousand thanks to the folks whose input made this project better: Kirstin Foot, Steven Livingston, Sandra Ristovska,

Monroe Price, Gina Neff, Phil Howard, Gary King, Abe Karem, Terence McDonnell, Jen Earl, Abdalrahman Ismail, Monther Etaky, Dan Gettinger, Hans Peter Schmit, John Krinsky, David Hess, Deana Rohlinger, Ian G. R. Shaw, Kevin Bales, Doreen Boyd, Stuart Marsh, Sandor Szöke, Mike Kobliska, Janine Schooley, Clara Eder, David O'Conner, Ben Tigner, Michael Strand, David Cortright, Claudia Martinez Mansel, Alex Hanna, Pat Meier, Joel Quirk, Ian Shaw, Matt Kravutske, Jon Bialecki, Akin Ünver, Dexter Pratt, Luis Michelsson, Josie Siegel, KATSU, Jesper Vestergaard, Cameron MacLeod, Adrian Borsa, Burrell Van Jr., Andrew Bergman, Toly Rinberg, and Eden Choi-Fitzpatrick.

The wonderful folks at the Hungarian independent journalism outfit *Atlatzo* helped get this project off the ground. I am particularly grateful to Tamas Bodoky, Áron Halász, and Akos Baranya. Critical early support was also provided by Eva Bognar at Central European University's Center for Media, Data and Society. Charlotte Lloyd helped me secure permissions from a number of places, including the Estate of Buckminster Fuller, Ruben Pater, Guy Wentborne, Pierre Bélanger, and KATSU. Gale Spitzley designed the cool "Observation Layers" figure in the second chapter, Kevin Dobyns tweaked all the graphs and figures, and Kristie Reilly copyedited the final document. Special thanks to my agent, Jill Marr at Dijkstra Agency, for her support in this process and Virginia Crossman for her fearless editorial oversight.

Final thanks to my partner, Joshua MacIvor-Andersen, and my spouse, Jenny Choi-Fitzpatrick. But really everything is Aila Pax Miyoung and Eden Justice Sunyoung—the world will be better in their hands.

I IDEAS

0 INTRODUCTION: TECHNOLOGY BEYOND SOCIAL MEDIA

This book started on the street.

It was nine o'clock in the evening and I was on the curb with a Hungarian police officer, who was asking for identification. Specifically, he was asking to see the papers of my graduate student, Tautvydas Juskauskas. In a former life, Tautis was a levelheaded lobbyist in his native Lithuania. In a future life, he would work for the world's largest drone manufacturer and later lead drone operations in Malawi for the United Nations Children's Fund. That evening, however, he was a suspect, wondering what he'd gotten himself into.

Tautis and I were in the process of documenting the largest street protests seen in Hungary since the collapse of the Berlin Wall. The government wanted to raise revenue by taxing the Internet traffic of every business and individual, whether at home or on a digital device. The officer was perplexed by our technology and by our role in the event. We explained that we were conducting research. He demanded our papers. We stalled (I'd forgotten to give Tautis the first lesson in Protest Fieldwork

101: Ditch the ID!), and finally I agreed to give the officer my name. I scribbled Austin Fitzpatrick, my legal name.

"Should we stop flying?" I asked the officer. He thought for a minute, looked at us, looked at our drone, shrugged, and waved us along.

The entire exchange lasted five minutes and drew a crowd of people, some of whom pulled out their mobile phones to document our conversation with the police. Perhaps the presence of citizen journalists bearing witness gave the officer pause. Perhaps he was going to let us go anyway. Whatever the case, we jogged off in an attempt to get ahead of the throng and set up our equipment in time to get aerial footage of the event.

As we arrived in the square, Tautis' phone rang. It was our contact at the local independent journalism shop. The crowd was almost there, he reported, and ready to engage the drone overhead. We'd planned what would happen next. The crowd was chanting together against the proposed tax, but also in defiance of the increasingly authoritarian government that proposed the law. With this momentum the crowd turned, as one, to point their phones upward. Together, they extended the decades'-old lighter sway familiar to any concert-goer into an entirely new space— pointed not toward a stage, but into the sky, directly toward our hovering drone. At the next protest we did the same thing, capturing the moment an even larger crowd poured over Budapest's picturesque Elisabeth Bridge.

It was this image that became iconic for the movement and that landed on the cover of the *International New York Times* the next morning. The point here is not that we documented a crowd, but that the *Times'* photo was of a crowd responding to our aerial

Egypt moves to carve out a buffer zone along Gaza

CAIRO

Homes are demolished as Sisi steps up response to attacks by militants

BY KAREEM FAHIM AND MERNA THOMAS

One day after an evacuation order, Egyptian Army bulldozers began demolishing houses along the border with Gaza on Wednesday, the first step in establishing what officials say will be a buffer zone intended to stop the passage of militants and weapons across the frontier.

The evacuations of hundreds of houses, mainly in the border town of Rafah, started on Tuesday and were part of a sweeping security response by the government of President Abdel Fattah el-Sisi in months of deadly militant attacks on Egyptian security personnel in the Sinai Peninsula, including the massacre of at least 31 soldiers last week.

That assault, on Friday, was the deadliest on the Egyptian military in years, and a blow to the government, which has claimed to be winning the battle against insurgents. Mr. Sisi, a former general, spoke of a "conspiracy" facing the state, though prosecutors have not yet identified any suspects.

Protest in Hungary Thousands of demonstrators held up mobile phones as they crossed the Danube in Budapest on Tuesday to urge the right-wing government of Prime Minister Viktor Orban to drop a proposed tax on Internet use. The protesters say the bill would choke off access to information not controlled by the government. PAGE 7

Fed declares victory with an end to bond buying

WASHINGTON

Program seen as success for U.S. economy may be hard for Europe to mimic

BY BINYAMIN APPELBAUM AND JACK EWING

An upbeat Federal Reserve, citing the strong performance of the United States economy, said on Wednesday that it was ending a bond-buying stimulus program widely credited with getting America back on its feet after the global financial crisis.

It is a program that the European Central Bank might like to emulate — if only it could.

The Fed said the bond-buying program had served its purpose by contributing to a stronger job market, which it said on Wednesday was continuing to improve. But seeking to reassure global financial markets, the Fed said it still planned to keep short-term interest rates near zero for a "considerable time."

The Fed's decision, which had been anticipated, was greeted with calm on Wall Street, with major indexes falling slightly after the announcement.

Of the big European economies, Britain has pursued its own version of the

Figure 0.1

Protestors point mobile phones at author's drone.

technology. Our drone didn't take the picture, it *made the picture possible*—it directed eyes and mobilized action (figure 0.1).[1]

The moment was both invigorating and symbolic. It was invigorating for the same reason like-minded people have engaged in collective action over the centuries: collective identity, collective effervescence, solidarity, and a desire to see things change. It was symbolic because it represented an early example of how new technologies enter public space and change politics in the process. Hungarian civil society groups had used social media sites to mobilize on the streets in real time against a threat to the Internet. Once on the street, they raised their digital devices toward a new witness to the entire affair: a small quadcopter that captured footage to be uploaded to the Internet the following morning, complete with a DJ Shadow soundtrack and a call to further action.

That event was one of several, as emboldened crowds saw in our videos something they didn't see in the local newspapers. Themselves.

The event was also symbolic because of what the crowd was protesting. Victor Orban, the Hungarian Prime Minister, had proposed a tax on the Internet that would have had a broad economic impact on all Hungarians, not just the usual gang of citizens who protested his anti-liberal agenda. As a result, grandmothers marched next to parents with strollers, and business owners and anarchists stood side by side, unified in denouncing his plan. The crowd underscored a point made by the sociologist Francesca Polletta: new technologies (and attendant public policies) create new reasons to protest.[2]

These protests attracted the international media. They also got the government's attention. In the face of this surprisingly strident display of solidarity and determination, Orban caved, the policy was abandoned, and the movement declared a victory. Later, Tautis did the math. He estimated that about 60,000 people took to the streets, virtually 3.5 percent of the city's 1.7 million residents.

While it may not sound like much, 3.5 percent is actually a magic number. Conflict theorists Erika Chenoweth and Maria Stephan contend that protestors' demands are met when they are both large and nonviolent.[3] How large do nonviolent protests need to be? Drawing on an impressive array of data, Chenoweth and Stephan suggest about 3.5 percent of a population on the streets, nonviolently, usually does the trick.

CIVIL SOCIETY TECHNOLOGY:
BEYOND SOCIAL MEDIA

I wrote this book out of fascination and frustration. Original fascination with our ability to support social movements on the street gave way to frustration with the lack of theoretical resources in social movement theory and the skepticism of some of our movement allies on the ground.

As a result, the core argument in this book is simple: technology matters for politics, and it matters in important but overlooked ways.

Our use of drones to document the size of protests is but one example of a growing wave of prosocial experimentation with new technology. In this volume, I focus on the way new tools are used by social movements, in particular, and civil society more broadly. Was there anything new about the way Tautis and I used our drone? Many great books have been written on the promise and peril of social media and the Internet. A fresh wave of thinking directs attention to wearable tech, artificial intelligence, and computational propaganda.

Our thinking about drones, in contrast, is a bit hazier, to say nothing of other technologies that lie beyond the new digital technologies of social media. This is a pity, as drones and other robots are showing up in all sorts of places.

But what are all these new devices doing? If you listen only to my good friends in the human rights world, the answers are chilling: drones represent a new wave of technology threatening civil liberties, violating privacy, and disrupting terrestrial approaches to security. New scholarship on these anti-social phenomena is growing at a rapid pace.[4] This book is about a quite different range of uses, with a lopsided emphasis on those that offer a clear public benefit.

That evening in Budapest left me with some nettlesome questions about how seriously we take this technology, so I spent the last few years gathering data on how drones are used, training civil society groups on the use of balloons and drones, collaborating with a research team at the University of Nottingham focused on the use of satellite data to document human rights violations, strapping GoPro cameras to 3D-printed gimbals on kites and balloons, and working with engineering and peace studies students to build and fly drones of their own.

Along the way it became clear that a whole spectrum of technology doesn't fit neatly into the contemporary conversation about "new media" and high-profile communication technologies like mobile phones. In my home field of social movement studies, we tend to focus on those moments when change agents identify things that are wrong with the status quo, frame those issues as problematic and change-worthy, then pressure those with power and authority to take action.[5] In countries like the United States this pressure can take the form of a boycott against a company that tests their products on animals or a campaign to pressure a politician to vote a certain way on environmental legislation. Growing attention is being paid to technology's role in these efforts. A boycott that was once facilitated by an important organization like Oxfam, Amnesty International, or Greenpeace might now be mobilized online and framed by a hashtag. Pressure on policy makers might have once come from a phone call, but can now come in the form of online campaigns and petitions. Scholars of politics, culture, and social change have spent considerable time exploring the impact of the new digital technologies that are critical to political communication.

We mustn't stop there.

In this book, I highlight technology's political impacts *before* and *beyond* social media.

The more time I spent reading about the role of technology in civil society, and particularly efforts focused on social and political change, the more I was struck by the dominant role social media plays in these narratives. Having joined many others in scraping Twitter data during the 2009 #iranelection, I was happy to see so much attention focused on the role of social media in mobilization and communication. Certainly, efforts to describe the utility of new digital technologies in political communication received a significant lift after the Arab Spring. I'm certainly not the first to focus on how civil society uses technology. My fellow travelers in social movement studies have spent some time thinking about the impact social media has on political mobilization, and their analysis of these relationships have hewed to one of four broad approaches.

The first is to ask fresh questions about technology's greater impact on the formation of collective identities in creating a sense of *we*-ness. People get involved online because they hope their voices will be heard, and see opportunities to come alongside others who feel the same.[6] Networks of websites helped create this feeling in a pre-social media era,[7] and properly configured digital spaces, including video gaming environments, can have the same effect.[8] New studies show that a younger generation of feminists found one another online, for example. As a result, networks grow and deepen, create community, and expand opportunities for future offline mobilization.[9]

The second is to emphasize the possibility that new technologies may help solve the kind of collective-action problems that have occupied near-continuous scholarly attention since the economist Mancur Olson first raised the puzzle in 1965: why do people choose to volunteer for social causes, especially if there's nothing "in it for them"? New studies emphasize the extent to which digital technologies lower the cost of coordinating and communicating.[10] Mobile phones, for example, help people spread information about important issues and help coordinate social and political action.[11]

A third approach to technology and politics emphasizes the extent to which engagement with technologies themselves create new political, economic, social, environmental, and legal realities. These include emergent online spaces for connection and collaboration around everyday projects, but also make room for politics, whether it's quotidian or disruptive.[12] New public practices are created by the routines that emerge organically out of persistent connective collaboration.[13] All of this engagement also creates new issues that themselves become sites of collective action—in other words, the Internet creates new reasons to get involved in contentious politics.[14]

The fourth broad approach to explaining the relationship between new digital technology and collective action is to call into question the findings of each of the three preceding clusters. Digital power is a two-way street, as states are often better equipped to harness innovation than are civil society actors. The result is a power disparity that distinctly disadvantages the public, or, even worse, sends civil society actors in the altogether wrong direction as they pursue *technological solutionism* and *Internet centrism*—an extreme version of the old adage *when all you have is a hammer, every problem looks like a nail.*[15] When all you have is a mobile phone, this line of thinking suggests, the solution to every social, political, or economic problem looks like an app. State and corporate power, in this light, is amplified rather than diminished by new techniques and technologies. Scholars of civil society ignore this broader fact at their peril.

If the benefits of quickly capturing and rapidly disseminating information were made clear in recent struggles for democracy in the Middle East and North Africa, the peril of poor quality control of this information has been on stark display in presumably settled democracies. This is clearly seen in the campaign that led to the election of Donald Trump in the United States and the attendant rise of algorithm politics and computational propaganda.[16] Not only does social media hold both promise and peril, we are now realizing, but its functions are underwritten by a range of invisible technologies—from algorithms to server farms—that are now the focus of public concern as well as scholarly analysis. I hope this book complements important new work from scholars exploring digital politics,[17] digitally enabled social change,[18] the logic of connective action,[19] hybrid media systems,[20] and the bits and atoms of technology,[21] each of which are discussed in greater detail in a theoretical afterword to this volume. I build on this prior work to advance a simple argument: technology plays a larger role in civil society than simply creating new social networks through social media.

Our experience in Budapest is a case in point. Tautis and I used a drone to generate a video released on Vimeo and shared on Facebook and Twitter. I had no trouble finding scholarship on the importance of networked publics[22] or rival advocacy networks[23]—that is, smart thinking about what we went on to do with the footage—but scholarship on the politics of the drone technology itself was harder to come by. What exactly are drones an example of? They are a new surveillance tool, clearly. But they are also an ideal platform for conducting citizen journalism and engaging in humanitarian interventions.[24] Drones are a new form of transportation infrastructure, but are also deployed as autonomous airborne Internet service platforms. Thinking narrowly about technology as a synonym for social media does not take us far if we need to think critically about tools that have so many different applications and implications. In this volume, a focus on communication is a necessary but insufficient condition if we want to understand the politics of technology and the technology of politics.

So what of technologies for social change?

Collective-action efforts in civil society rely on key organizational and infrastructural capacity. These efforts also include a growing constellation of tools for gathering and analyzing data—both bits *and* atoms matter to civil society.[25] This fact requires a broader way of thinking about the relationship between civil society and technology. The first step is to better situate the role of communication within a larger technological landscape. Message creation, reception, and interpretation are not the be-all and end-all of technology. We also use tools to warehouse and transmit data, for example.[26]

This is important to note, as the amount of web traffic between machines outpaces the volume sent or received by humans, meaning computers are talking to one another at higher rates than humans are. Within my home domain of social movements, communication is important for gaining public acceptance and raising the cost of the status quo, as the case of naming and shaming a corporation clearly demonstrates. When social movements want to make a difference, they create posters, websites, and hashtags to communicate their demands. But they also do whatever it takes to make old practices too expensive to maintain. This is the logic behind lunch-counter sit-ins during the Civil Rights era, die-ins during protests against the US invasion of Iraq and Afghanistan, and the creation of barricades to thwart authorities during the Paris Uprising, romanticized in *Les Misérables* in the figure of Inspector Javert.

This book unpacks the relationship between these struggles and technology writ large. Such a project requires thinking in much broader terms about what counts as technology. It also requires importing some concepts from distant scholarly lands. To be blunt, research for this book pushed me out of my academic comfort zone and into a bit of a walkabout with folks thinking about infrastructure, art, architecture, the history of science and technology, human-centered design, ethics, and engineering—all the while doing my best to remain focused on what I really care about, which is how technologies shift the balance of power—however modestly or temporarily—in favor of the people.

The focal point, then, is that interplay between technologies and civil society, as well as its impact on politics. By *civil society* I simply mean those *activities, institutions, and spaces that are separate and independent of both the state and the market*. Such a broad approach turns our attention to the political life of technology, since new tools change what we think of as public space—creating new public spheres. Several of the cases in this volume represent *collective challenges to systems or structures of authority*, as clear a definition of social movement as there ever was.[27]

I wrote this book out of a conviction that technology plays a larger role in advocacy and social change than just capturing and distributing moving images. In my last book, *What Slaveholders Think*, I explored the impact human rights campaigns have on feudal socio-economic relationships. In particular, I focused on how contemporary slaveholders respond to grassroots challenges to their authority. The stories that people told me were rich with technologies of all sorts—mobile phones alerted once-disconnected workers to new opportunities in nearby cities, new stone crushers reduced demand for workers, and farm implements were occasionally used in violent uprisings. New digital technologies matter, but so do technologies that are decidedly old, or that have nothing to do with communication. A few years back my partner—an international aid worker—was a few days late in her arrival to Tanzania, where we lived at the time. She had been delayed in Ethiopia's hinterlands by anti-government protestors who had blocked the road with large boulders. In that particular context, a blocked road was the tactic preferred by those struggling for change. Half a world away, an American collective called Public Lab sells basic science kits so grassroots environmental activists can gather their own data about environmental and health conditions in their area. This citizen science is a form of public engagement that challenges the hegemonic grip official science has on the gathering and interpretation of facts.

What follows is my attempt to take each of these technologies seriously. What do a drone in Budapest, a new stone crusher in rural India, boulders in Ethiopia, and a petri dish in Flint, Michigan, have to do with one another? In this book, I suggest one possible answer: collective-action efforts use tools and technologies to get their jobs done, and this use and those tools are far broader than anticipated by a narrow focus on *new digital technology*.

In the next chapter, I will suggest that the notion of a public sphere allows us to better recognize civil society's important connective components while also pushing out how we consider technologies' spatial implications. Digital technology has democratized important social, political, and economic activity, enabling individuals or groups to do things that had previously been the remit of states and large corporations.[28] For example, a host of tools for citizen science are available at PublicLab.org, allowing people from all walks of life to test their air and water, and in so doing to produce counter-hegemonic data about the environment. While it is beyond the purview of this study, science itself is a site of contentious knowledge.[29] Social change efforts regularly draw on technology. They always have. *New digital technologies* play an important role in politics, but are only one piece of a much larger puzzle.

TOOLS AND TECHNOLOGIES

Social change advocates use technology to raise awareness and connect people. Technology is also used to make the status quo too expensive for movement targets, or to gather and analyze data on important social, political, economic, or environmental events, or to simply catalog and archive raw data for future analysis. If we want to better understand and document the way technology gets used for politics, then we must start with a clear and scalable definition of technology.

In this book, I adopt a simple conceptualization of tools and technology as *objects in use,* whether they are digital or analog, physical or virtual, or used by human or nonhuman folk.[30] A focus on use is meant to direct our attention to the humble everydayness of the tools that comprise so many of our collective efforts, as well as the importance of protest slogans, codes, algorithms, bugs, and viruses (both biological and digital). This definition includes participatory objects, settings, devices, and other "stuff" that acquire explicit political capabilities through their use.[31] Technology—what Langdon Winner has called "all practical artifice"[32]—is stuff that becomes politically, socially, personally, and economically useful when we put it to use.[33]

Do we need complicated definitions in order to understand the world around us? Most readers know that the answer is *no*, but academic readers might need a bit more convincing. In recent social science, "the overwhelming focus has been on texts, the industry that produces them, and the viewers that consume them. As a result, the materiality of [media] devices and networks has been consistently overlooked. ... The headlines are examined but not the newsboys who shout them, the teletypes that clatter them out, or the code that now renders them into clickable hyperlinks."[34] This critique, leveled against media scholarship by Tarleton Gillespie, Pablo Boczkowski, and Kirsten Foot, is broadly relevant, as technological changes require better theories.[35] My goal in this volume is to remedy this oversight by pointing to technologies before and beyond social media.

I have clustered each of the following chapters according to the work I believe them to be doing. Chapters 1 and 2 introduce the book's key ideas. Chapter 1 defines key concepts and advances a number of testable hypotheses relating to technology's "emergent and disruptive" use. This chapter positions the entire project within broader academic conversations, and the non-academic reader may be advised to skip the first chapter entirely. The second chapter provides a sociopolitical history of the kite, the balloon, the satellite, and the drone, arguing these "geospatial affordances" have played a role in expanding the public's capacity to bear witness to important issues ignored by states and markets.

Chapters 3, 4, and 5 detail a number of interventions that these ideas make possible. Chapter 3 provides case studies of drone use, exploring whether they are emergent or disruptive, and in this way tests some of the book's key ideas. Chapter 4 responds to the book's central call for studies that take the politics of technology seriously, and in a case study of the camera—*rather than the image*—emphasizes a few of the puzzles that emerge from a focus on technology's politics. Chapter 5 explores how drones are used to resist the status quo, while surveying the legal and physical techniques used to resist drones.

The final section of the volume focuses on the implications of these ideas and interventions. In this sixth chapter, I extend some of the book's key arguments to a broader range of cases, in this way testing their portability to other contexts. Perhaps helpfully, the volume also includes a theoretical afterword, in which I take a deeper dive into some of the academic conversations this book draws from and contributes to.

I hope to leave the reader with a strong sense that technologies play a broader role in social and political struggle than is indicated in studies that focus on Twitter, the Internet, and Wikileaks. Those are important, but they are part of a larger constellation. It is to that broader space we now turn.

1 TECHNOLOGY FOR THE PUBLIC GOOD: EMERGENT AND DISRUPTIVE

What happens when we take materiality seriously? That's the question asked by a fresh wave of scholarship on the impact that physical stuff has on social relations and social action, a trend seen in the titles of recent books such as *States of Knowledge*,[1] *Human-Built World*,[2] *New Materialisms*,[3] *Materiality and Organizing*,[4] *Bits and Atoms*,[5] *Signal Traffic*,[6] *The Undersea Network*,[7] and *Stuff of Bits*.[8] Some conceptual looseness is needed to recognize the complexity of including software, algorithms, and artificial intelligence into our definition of technology alongside websites, zines, and barricades. I am partial, for example, to the Armenian term for material—նյութական—which translates to both *corporeal* and also *ponderable*. Thinking about *tools in use* points us back to the "tightly-interwoven relationship between the material and the symbolic" that technology often represents.[9]

Communication is one important use, and communication technologies have long proven critical to collective-action efforts. Yet important and high-visibility technologies like the printing press, radio, telephone, television, Internet, and mobile devices are but the most visible islands in vast undersea ranges that shape currents flowing nearer the surface. New scholarship on digital infrastructure, undersea cables, artificial intelligence, political bots, the Internet of Things, smart cities, and wearable technologies are broadening our understanding of what counts. Much can be learned from a focus on the ways non-state actors adopt technology and technological capacities more broadly, in digital *and* analog forms of communication and beyond.

Activists and advocacy groups use tools to draw attention to their cause and mobilize support, but they also use tools and techniques to generate political leverage by making the status quo too expensive to maintain. And advocates often use technology in ways the public never sees, and might generate data that is never entirely communicated to the outside world—government accountability groups do this on a regular basis. The Environmental Data and Governance Initiative and the Sunlight Foundation spearhead initiatives to archive and monitor government web resources, including critical information stored on the websites of federal agencies like the Environmental Protection Agency and the Department of Health and Human Services. This *data resistance* came in response to a widespread and credible fear that the 2016 election of Donald Trump would threaten evidence-based policy making.[10]

A host of material artifacts and objects are also used to do things like protect political challengers from the elements (umbrellas, clothes, and offices) and make life difficult for incumbents (barricades, slashed tires, and denial of service attacks online). The tactical tools most readily identified with police clashes prioritize practical issues (preventing tear gas from entering one's eyes and lungs) over the communicative capacity of those materials. Other tools, including paste, poster boards, and spray paint, must be recombined in order for their communicative potential to emerge as posters or graffiti. Online tools, including Tor, political bots, and viruses, are not necessarily geared toward communication, but may be flexibly combined and deployed for any number of objectives.

Such a list is pragmatic rather than normative, as many tools can be used for either violent or nonviolent purposes and may be used by nonprofits, community-based organizations, nongovernmental organizations, social movement organizations, the communities they serve, or the incumbents they target. An approach emphasizing objects and their use also includes complex technologies that are themselves able to produce tools, as when a printing press produces an event flier or when a 3D printer produces a printing press.

Each community of users creates unique logics of use. Advocates of social change have long debated the utility (rather than the ethics) of violent tactics. Some have suggested disruptive protests are more likely to secure important gains,[11] while recent empirical work suggests the evidence is with Mahatma Gandhi and Martin Luther King Jr. on this count.[12] As a result, some tools cluster around particular normative or tactical commitment, like nonviolence. Other patterns of use emerge in time- and space-bound ways and reflect local resources and folkways. For example, barricades and balaclavas are important tools in urban protest, but everyday farming material—seeds, hoes, soil—may comprise the tools of the rural weak, as James Scott has so memorably demonstrated.[13] Of course, norms and materials intersect in many ways—farmstuff like fertilizer and construction material like nails can be deployed as an improvised explosive device, but violent tactics are rejected by most change-oriented advocacy efforts in settled democracies. These configurations depend on time, place, and resource.

How should we think about these tools and their use? It's this chapter's job to answer that question, though at this point I feel obligated to be frank with the reader. The rest of this chapter has been written in the hope of better connecting two important scholarly communities. As a result, it strays from time to time into technical details that are intended as a bit of note-passing between my colleagues focused on technology, media, and society and those focused on contentious politics, protests, revolutions, and social movements. Readers interested in how drones, satellites, balloons, and kites are used, and who would rather not read about repertoires, affordances, and whether nonhuman living beings have agency, might rather skip ahead to the next chapter. For those who continue: don't say you weren't warned.[14]

REPERTOIRES AS CLUSTERS OF THINGS IN USE

Thus far I have argued that civil society actors use tools for many purposes, including awareness-raising (i.e., communication), data storage and analysis, and the creation of political leverage through obstruction and cost-raising. Technology is regularly used to gather and analyze data crucial for decision-making within organizations or concession-extracting from powerful institutions. While informing the public and mobilizing constituents are critical ingredients in the politics of social change, they are not the only way technology is used. The term *repertoire* has been used by social scientist Charles Tilly to describe the broad constellations of strategies and tactics used to encourage or thwart social and political change.[15]

I propose we can apply this logic to tools in use, such that a technological repertoire is simply the *broad and repeated use of tools and techniques.* This framework may help us to better conceptualize and debate technology as a field of action or a state of play, rather than as a fixed and stable inventory of stuff that's just sitting there.

A dynamic approach is best, since repertoires are not inherently stable. They are instead nested within broader contexts and subject to spurts of human creativity or the drag of precedent. Repertoires are part of the status quo, and the status quo is almost always being consolidated or challenged. The present moment—any present moment—is only a set of settlements. All *longues durées* are way stations.[16]

Every field of action has its own repertoire and every community and every struggle has its own way of doing things. Intergovernmental organizations are more likely to rely on a host of tools and technologies that facilitate state-level coordination and communication. These could include security protocols, specialized communication channels, and centralized headquarters. The United Nations offers a perfect example of these factors. The UN's white vehicles, blue helmets, and branded supplies are all part of their material footprint. The UN ecosystem relies on the technologies of bureaucracy, like modern office systems, as well as tools of logistics and infrastructure, like tarps, radios, and shipping containers.

Nongovernmental organizations rely on technologies that run the gamut from generic office systems to purpose-built technologies in the field. Public-service campaigns focused on reducing the prevalence of disease, for example, may distribute bed nets, vaccines, and condoms. Sociologist Terrance McDonnell has documented the extent to which the aid industry relies on cultural objects over which they have very little control.[17] For example, the female condom creatively doubles as a bracelet, and mosquito nets are used to sift sand or are repurposed as wedding veils.[18] While it is beyond the scope of this volume to document the number of ways nongovernmental organizations use technology, it is safe to say that such an inventory would find a staggering range of practices, as means and as ends, across context and over time.

Community-based organizations are groups that prioritize a grassroots connection to a particular place or group. Since they rely on local support, their legitimacy is critical. As a result, organizational form, leadership composition, and the nature of material resources are all subject to local considerations. Community-based organizing, especially, requires buy-in from the community, as these efforts often rely on a theory of power and social change that prioritizes the role of local voices and experience in creating bonds of solidarity that allow for broader impacts. Getting things wrong in these contexts can mean the difference between authenticity and a perception that a group or person is fake.[19] Here two things are on display. The first is the power of the repertoire, and the second is the power of community efforts that get the repertoire right. As a result, it is difficult to simply drag and drop tools or technologies into communities and expect buy-in. This is another of the key takeaways from important sociological analysis of the humanitarian aid industry—people have agency and use aid material as means to suit their ends. In my own work, I have documented a village of bonded laborers in rural India that mobilized against the upper-caste landlords who were working them to death in their stone quarries. Enraged, the workers picked up the resources at hand—rocks—and pelted their abusers.[20] A particularly abusive member of the landlord class was killed in the ensuing violence.

Repertoires emerge from local material, economic, political, and social conditions.

For social movement organizations, repertories describe the cluster of movement tactics that are available and desirable at any particular point in time. Again, we find no comprehensive list, but rather a rolling constellation of approaches. In the 1960s, activists in the New Left used a range of tactics, including "petitioning, rock throwing, canvassing, letter writing, vigils, sit-ins, freedom rides, lobbying, arson, draft resistance, assault, hair growing, non-violent civil disobedience, operating a free store, rioting, confrontations with cops, consciousness raising, screaming obscenities, singing, hurling shit, marching, raising a clenched fist, bodily assault, tax refusal, guerrilla theater, campaigning, looting, sniping, living theater, rallies, smoking pot, destroying draft records, blowing up ROTC buildings, court trials, murder, immolation, strikes, and writing various manifestos or platforms."[21]

Writing in the same era, nonviolence advocate Gene Sharp proposed a list of 198 nonviolent tactics, including skywriting and earthwriting, protest disrobings, and the destruction of one's own property.[22] Of course, not every struggle for social change uses even a fraction of these tactics. Nevertheless, the list points to the range of nonviolent options available within the broader repertoire of change-oriented approaches on offer at one time (the 1960s) and in one place (the United States). These broad strategic goals included raising awareness, changing public opinion, and raising the cost of the status quo. The goal is usually to secure public support and force official action on new policies and legislation related to rights, resources, and recognition. There is nothing about change-oriented strategies that prescribes particular tactics, and much debate continues about particular constellations of tactics, as evidenced by the ongoing debate over the utility of violence.[23]

Repertoires are appropriate in different times and places, and they are also subject to cultural and material constraints. Social actors that rely on public approval cannot adopt approaches that are in opposition to key values and beliefs within their host society, nor can they engage in activities that are unintelligible in the local idiom.[24] A nonprofit organization working to end animal cruelty would have a hard time justifying a fundraiser in which a hunting safari was on the auction block, for example. Whether raising money in a fundraiser or raising hell on the streets, even the most creative actors must draw on or innovate around the material in their immediate vicinity, whether it be a website or a boulder.

Making demands effectively, Charles Tilly argues, depends on people and groups having "a recognizable relation to their setting, to relations between the parties, and to previous uses of the claim-making form."[25] The same logic applies to the use of technologies. It is not technological determinism to observe that cobblestones are handy missiles in a clash with police, or that a protestor cannot throw cobblestones at the authorities if they have been preventively glued to the ground.[26] The mass circulation of pamphlets was impossible before the invention of the printing press, yet an anthropomorphic bias directs our attention *away* from the printing press and toward the pamphlet getting read by publics who then gather in front of Parliament.[27] This bias is understandable—politics are often by and for human groups—but an eye toward hybridity suggests both the technological and the political are important, something scholars like Benedict Anderson have gotten right in describing the importance of print media.[28]

A pattern is evident: in our accounts of advocacy and social change, the means of production are overshadowed by the modes of engagement. This difference in perspective—was it the printing press or the public?—is important, as technology is both enabling and constricting of collective action, both prior to and simultaneous with advocacy efforts. Technology is causal at the structural level—it lays the groundwork for action—and critical at the level of the lived experience of social actors. Particular repertoires draw on material culture and involve the use of placards, fliers, and other objects and artifacts, making up an analog and urban protest repertoire familiar from the nineteenth century to the present.

Simply put, individuals and organizations interested in social change need tools and technology to perform whatever tasks are appropriate within their particular contexts. They do so in ways that match both public opinions and organizational philosophies. Over time, with neither planning nor intent, these efforts and actions aggregate with other users' uses to produce repertoires of action.

Repertoires are emergent, rather than stable, fixed, or predictable.

AFFORDANCES: WHERE TOOLS COME FROM

At this point, we begin to see the broader argument as it stands: Technologies are really just things that get used, and over time usage clusters into patterns that I've called repertoires. Yet how does an assessment of usefulness come about? Is it a function of the object or a function of the actor? Do some objects have their own purposes (realism), or are all objects simply social constructs (constructivism)? This debate has been central to science and technology studies for the past two decades, but is seen less often in efforts to understand politics and social change.

Those of us trained in the social sciences are better attuned, unsurprisingly, to the ebb and flow of political and social processes than we are to technology's currents. We would do well to pay a bit more attention to *affordances*,[29] which simply refers to the *possibilities that things offer for action*.[30] The possibilities for action differ by time and place, but they also differ based on a range of features unique to the human actor.

Humans have different intentions for an object's use, and not everybody imagines using things in the same way.[31] As Winston Churchill famously argued, "we shape our buildings; thereafter they shape us."[32]

This is essentially the claim made by a line of scholarship on the social construction of technology. It is human action, not technology, that shapes how tools are identified and used. In a seminal article on technologies, tools, and technological artifacts, the sociologist Ian Hutchby adds nuance to this approach, suggesting tools "do not amount simply to what their users make of them; what is made of them is accomplished in the interface between human aims and the artifact's affordances."[33] Technology is more than the incubator of new forms of social relations, since "social processes and the 'properties' of technological artifacts are interrelated and intertwined" in theoretically important ways.[34] Here we see the outlines of a world comprised of human and nonhuman actors operating in a broader network, within which all have agency—as in the actor-network theory developed by French philosopher Bruno Latour—and people as well as material objects are *actants* with agency sufficient to shape our lives.[35] Speed bumps are the classic example of how the actor-network approach conceptualizes the agency of artifacts. They are as vital of an actor, in Latour's thinking, in the flow of suburban life as any human; shaping—demanding, forcing—a human response.

Not so fast, argue scholars Wiebe Bijker and John Law:

> Technologies do not have a momentum of their own at the outset that allows them...to pass through a neutral social medium. Rather, they are subject to contingency as they pass from figurative hand to hand, and so are shaped and reshaped. Sometimes they disappear altogether; no-one felt moved, or was obliged, to pass them on. At other times they take novel forms, or are subverted by users to be employed in ways quite different from those for which they were originally intended.[36]

Female condoms doubling as bracelets and fishing nets used for wedding veils make this point quite neatly. The human is in charge. Speed bumps, in this light, are designed, installed, driven over, and eventually replaced by humans, the ultimate agents.

Recent thinking about affordances for advocacy and social change has directed attention to digital tools. Here the concept of *leveraged affordance* suggests people take action and use things for their own ends, regardless of what the thing itself was made for. There is general agreement in this space. Sociologists Jennifer Earl and Katrina Kimport suggest digital affordances are a "type of action or a characteristic of actions that a technology enables through its design,"[37] and political communication scholars Lance Bennett and Alexandra Segerberg argue interactive affordances facilitate political engagement and provide broad opportunities for action.[38]

How does the concept of affordance help us think about the work Tautis and I have done with drones and balloons? The answer comes from media scholar Steven Livingston, who has suggested that digital affordances come in three types: 1) digitally networked affordances; 2) forensic affordances; and 3) geospatial affordances. It is the latter, Livingston argues, that provide the ability to flexibly deploy "spatial and panoptical awareness and virtual presence."[39] Livingston's approach takes an important step in shifting attention from new digital technologies like the Internet and social media to a broader constellation of tools for sensing and seeing the world. Livingston's own work focuses on satellites, but the lesson resonates more broadly: we have an opportunity to recognize a wider array of tools and objects and to debate the range of ways they are used.

Opportunities for action don't just sit on the shelf, patiently waiting to be used. Rather, opportunities must be recognized as opportunities. This may seem obvious for most readers, but students of contentious politics have long debated two critical puzzles. The first is whether opportunities and threats can be identified independent of the event they are trying to explain—in other words, can the right moment for political action be reliably and independently identified by an objective bystander? The second question, more pressing for our work here, is whether the impetus for action lies within a political moment and broader environment, or is in the hands of individual agents who make their own history. The stakes are high, as these questions get at the heart of how social change is thought to happen.[40]

Scholars focused on materiality, Latour and Hutchby among them, offer direct responses to these concerns: potential exists independent of perception. Speed bumps bump whether or not I notice them, and regardless of whether I brake. Furthermore, while the particular features of a physical object set limits on what *can* be done with an artifact, they do not constrain the human agent's range of experimentation with what *might* be attempted.[41] In fact, it is through experimentation itself that new innovation happens and discoveries are made. An affordance may be false, but this might not stop a human agent from using it anyway for whatever end they have imagined.[42]

I like Ian Hutchby's approach, because he charts his own path, rejecting both realist accounts of technological affordances, which suggest that objects have "inherent properties that act as constraints," as well as constructivism, which considers the reality of these same objects to be the result of "discursive practices in relation to the object."[43] It is in their actual consideration and use by human agents—that is, in relation—that potential emerges.[44] This potential resides in both the worldly object and the human agent. A complementary argument is made by sociologist Gina Neff, who suggests technological determinism is a red herring, and rejecting it out of hand avoids important and unresolved questions about how tools are designed, how tools function, and users' awareness of the power and position of tools. Neff suggests that a narrow focus on a few examples in actor-network theory (like Latour's agentic speed bump—we know it has agency because it makes us slow down) ignores the importance of *scale* and *scope*.[45]

By this logic, it is possible to produce a number of hypotheses: in the short run, humans have agency to create and use tools; in the medium run, institutions (groups of humans) have agency to further shape the institutions and systems that govern tool use; in the long run, it is systems that shape human action and define the context in which individual agency is exercised; and each of these factors is in play simultaneously in every society for a wide range of processes, none of which are as linear as these hypotheses suggest. This observation is not new, and in fact simply incorporates tools into more traditional assessments of social change.

My focus here is on the first two of these hypothesized stages, as humans and groups of humans develop and adopt technologies, with particular attention to how they are used, and the implications that flow on from there. The significance of an object or artifact remains so long as it is used—passed from figurative hand to hand, to use Bijker and Law's metaphor. Presumably, tools are used, revised, reimagined, and passed along because they fit comfortably within the broader political, cultural, and economic landscape. They can be seen, can be acquired at a reasonable cost of energy or money, are useful for some purpose, and seem appropriate considering the norms and values of the society people find themselves in. When tools are used often enough and are shown to meet key objectives, a technological repertoire emerges.

But what of politics?

Early decisions about technology, political scientist Langdon Winner suggests, are actually decisions about politics and, even more fundamentally, are about what we value as a society. More important than a focus on user intent is the recognition that "the technological deck has been stacked in advance to favor certain social interests and that some people were bound to receive a better hand than others."[46] Humans make decisions about technologies that go on to have causal impacts on humans, Winner argues. In other words, affordances are "the dynamic link between subjects and objects within sociotechnical systems";[47] they are one of the stages on which the stuff of politics—interests, grievances, hopes and fears—takes form and takes flight.

Humans choose to use tools for all sorts of reasons. It is important to ask, alongside scholars like Ian Hutchby and Gina Neff, how and why this happens. It is also important to explore the implications of these choices. Cautious skeptics like Langdon Winner and caustic critics like Evengy Morozov are both at pains to emphasize the fact that technology can build power or erode power, and whether this power accrues to the already powerful or to those struggling for justice should be a matter of grand public debate. Social actors use affordances they think will help them accomplish their objectives, and this use is nested within existing circuits of power. Affordances are the emergent property of material and perception, of objects and actors. They are therefore always in flux, as broader social, political, and economic contexts shape both what exists and what can be imagined. These broader contexts also rest on systems of production that prepare various admixtures of raw material for consideration.

AGENCY IN AN ERA OF ARTIFICIAL INTELLIGENCE

The story I have told thus far is that scholars of politics and society should take technologies more seriously, since institutions and individuals already do; that clusters of usage form patterns and habits (repertoires); and that this usage is shaped by a sense that particular tools are useful. In so doing, I have struck a middle ground between realist and constructivist approaches to the material world, settling for an assessment that recognizes the capacity inherent in physical objects, independent of perception, but ultimately privileges human agency in the deployment of these objects.

Will this always be the case?

The simple answer is that, no, this will not always be the case. Rapid developments in artificial intelligence are adding new wrinkles to this story. If change is coming, what should we be on the lookout for? Answering this question requires a more nuanced understanding of the nature of agency. In their landmark text *Acting with Technology*, Victor Kaptelinin and Bonnie Nardi suggest agency is the ability and need to act in such a way that produces effects (for one's self or for others), and is a "fundamental feature of both the subject and the object" of any particular *interaction*.[48] An emphasis on both *ability* and *need* suggests that, while both people and things are regularly "acting-in-the-world," their agency is not all of the same sort. Networks of interaction, Kaptelinin and Nardi argue, have *asymmetric degrees of agency*,[49] a direct challenge to Bruno Latour and actor-network theory's more generous and homogeneous consideration of a blanket agency that describes the impact of both human and non-human actors, but also a tacit admission that some things make demands and thus have the effect of slowing us down.[50]

Kaptelinin and Nardi account for multiple agencies through a typology that recognizes biological and human needs ("needs-based agency"), action on someone else's behalf ("delegated agency"), and unintended consequences ("conditional agency").

A need to act, they argue, can come from either biological or cultural needs.[51]

Needs, then, are something animals have, but speed bumps don't.

This approach can be found in table 1.1, which originally appeared in Kaptelinin and Nardi, but has been reproduced (and modified) here. *Things* can be natural nonhuman entities, or they may be cultural bits of the built world. In either case, they produce effects, though not because they have needs of any sort. Likewise, nonhuman living beings produce effects and act according to their own biological needs, but these also come in two forms. The first is natural and independent of humans. The second is cultural and consists of the world that humans have built for themselves, including domestic animals, plants and fungi, live vaccines, and clones. The things and nonhuman living beings that are part of the human-built world exercise *delegated agency*, in that they realize the intentions of other human beings. Humans, for their part, hit on every note, since they produce effects, act according to biological and cultural needs, and are able to manifest the intentions of others, or to resist doing so. Finally, Kaptelinin and Nardi argue, social entities like the United Nations produce effects, act on cultural needs, and realize the intentions of others, but have no biological needs of their own.

Readers who came to this book out of an interest in technology and social change are right to ask what all of this has to do with the case at hand. My response is found in the shaded column in table 1.1 above, which I have taken the liberty of adding to Kaptelinin and Nardi's original six-category set of actors. This book is written at what may be the dawn of an era in which new sets of autonomous devices and systems will require new think-

Table 1.1 Kaptelinin and Nardi Agency Typology (modified), 2006

Agencies	Agents	Things (natural)	Things (cultural)	Nonhuman living beings	Nonhuman living beings (cultural)	Nonhuman living beings (emergent)	Human beings	Social entities
	Examples	Tsunamis, Northern lights, vernal pools, Martian rocks	Speed bumps, sewing machines, teapots, adzes	Grizzly bears, California poppies, truffles, protozoa	House cats, Dolly the sheep, GMO corn, Bourbon roses	Artificial intelligence, deep neural nets	Spinuzzi's traffic engineers, Miettinen's, scientists, ANT's princes	World Trade Organization, ISO, Doctors Without Borders, United Nations
Conditional agency	Produce effects	+	+	+	+	+	+	+
Need-based agency	Act according to own biological needs	−	−	+	+	−	+	−
	Act according to cultural needs	−	−	−	−	−	+	+
Delegated agency	Realize intentions of (other) human beings	−	+	−	+	+	+	+

Note: For a detailed exploration of the original table, see Victor Kaptelinin and Bonnie A. Nardi, *Acting with Technology: Activity Theory and Interaction Design* (Cambridge, MA: MIT Press, 2006).

ing about what, exactly, constitutes *need*. I am neither technically equipped nor intellectually prepared to argue that artificial intelligences are on track to develop their own biological and cultural needs, but enough is known about emergent properties in complex systems to anticipate that an era of radically unpredictable sociotechnical change may lie before us, especially in the form of self-healing within large algorithms and artificial intelligence— two breakthroughs that drones rely on, for example.[52] This fact will be relevant to both scholarly debates over structure and agency as well as practical efforts to shape the world through contentious politics.

This additional column creates room for the emergence of what Kaptelinin and Nardi themselves anticipate: "actual artifacts with intentions or desires" may emerge from innovation and advancement in artificial intelligence.[53] To imagine how we might get there, one need only imagine a drone tasked with hovering over a particular place, pausing to recharge its batteries. Is recharging batteries a need? Are survival and energy renewal different things, and if so, how? This idea is most provocatively explored by Nick Bostrom, whose book *Superintelligence* imagines the apocalypse that would follow if an artificial intelligence was tasked with the apparently simple job of collecting paperclips. Radical and focused attention to this simple imperative could bend all knowledge, resource, and production to this effort, and in this way erode or destroy all the other atoms and bits of the world that make life livable.[54]

I began this book in the hopes of better understanding and explaining several near-future applications for the kind of drones being used by both police and protestors. These devices have become safer to fly thanks to a sophisticated combination of sensor arrays and control systems, the very features that allow them to autonomously enter and navigate environments, gather data, and, increasingly, to take action on that data. Such a combination of mobility, sensing, and action capacities suggests the possibility of emergent activity beyond the original intentions of the programmer responsible for the algorithm.[55]

This is a challenge that I bravely leave to others, but that recurs throughout this volume, as well as in debates about drone warfare and autonomous weapons systems more broadly.[56]

EMERGENT AND DISRUPTIVE TECHNOLOGY

Technology can change the balance of power and help hold the powerful to account. These uses are usually thought of as disruptive. One look at the headlines confirms why the concept *disruptive new technology* is popular. Yet the phrase is more frequently used than explained, and much good could come from a clearer articulation of what we mean by it. Technologies, as I use the term, are *tools in action*. But what are we to make of the concepts *disruptive* and *new*? At the broadest level, I would like to use the term *disruptive* to signal the use of a tool that is politically or socially unacceptable and does not jibe with dominant repertoires at play in a particular social, economic, or political context. More parsimoniously, it is the use of a technology whose means or ends enjoy little initial approval. Likewise, the most parsimonious approach to *emergence* simply indicates whether a particular task can be accomplished with current technology.

Where disruption asks *should*, emergence asks *could*. The simplest version of this argument is found in table 1.2.

Immediately, we must add caveats.

Table 1.2   A Primitive Typology of Emergent and Disruptive Tools

		Emergence	
		Can be done with current tools (non-emergent)	Cannot be done with current tools (emergent)
Disruption	Follows norms (non-disruptive)	*Definition:* Ends can be reached with current technological means Broad approval for ends or means	*Definition:* Ends cannot be reached with current technological means Broad approval for ends or means
	Challenges norms (disruptive)	*Definition:* Ends can be reached with current tools Little approval for ends or means	*Definition:* Ends cannot be reached with current tools Little approval for ends or means

This book is full of instances in which non-state groups use drones to accomplish tasks that had previously been impossible. Yet powerful nation states have historically been able to accomplish these same tasks using large and expensive technology like helicopters. Practically, in the pages that follow, I use the term *emergent* to ask: *is the device performing a task that can be performed by civil society using other tools?* If the answer is *yes,* then the tool is not new and therefore not emergent. Observant readers will note that in narrowing my inquiry to civil society actors, I am deliberately excluding those tools available to powerful actors like states and corporations. A higher threshold—physical impossibility of task performance by other means—should be taken seriously, and indeed I highlight a few such cases in the pages that follow. Others may develop more particular understandings.[57]

Under a strong interpretation, requiring physical novelty, any task that can be performed by an airplane or helicopter is not new. Under a weak interpretation, any task that cannot be accomplished by a moderate-sized civil society group working to influence the state or market is new. I leave it to the reader to adjudicate between these two thresholds, the former being political and the latter being technological and economic. The observant reader will note that I am avoiding the question at its broadest, which is whether, as Winner asks, "modern technologies added *fundamentally new* activities to the range of things human beings do."[58] This question was first raised about three thousand years ago by Solomon, and I leave its answer to the reader.

Determining whether a technology's use is disruptive requires us to ask: is the device performing a task that may be *acceptably* performed by civil society using other tools? In other words, is the usage sanctioned by the dominant social and political norms of the day? Political norms are often shaped by the expectations that people have for accountability, authority, and control. Cultural norms are often shaped by our expectations around privacy, safety, accountability, transparency, and the interplay of these factors.[59] These norms vary between countries, regions, polities, and cultures. It should be no surprise that the use of drones would be governed by public policies that emerge at the intersection of political pressures and cultural norms. In some countries, cultural and political norms may conflict, as when the public demands levels of transparency and accountability that the political establishment is unwilling to accommodate.[60]

Readers unhappy with my terms are invited to find others, as my goal is not to create cumbersome new taxonomies or start tendentious academic debates, but to instead disconnect phrases like *disruptive new technology* from an axiomatic association with digital tools like social media and the Internet. In the pages that follow I'll be arguing that the phrase *disruptive new technologies* describes, at varying times, kites, balloons, satellites, and drones. It is to these technologies that we now turn.

2 DEMOCRATIZING SURVEILLANCE: DRONES, SATELLITES, AND BALLOONS FOR THE PUBLIC GOOD

I wrote much of this book from my office in San Diego, an hour's drive from the birthplace of America's military drone industry. The birthplace is a garage—standard-issue R&D space for California innovators. The garage belonged to Abraham Karem. Born in Baghdad in 1937 to Jewish parents, Karem immigrated to Israel in 1951, where he worked as a prolific engineer. He built his first drone during the 1973 Yom Kippur War, at a point when Israeli engineers were moving from ideas to field tests in six months—an unheard-of rate of innovation. He brought this experience to the United States in 1977, and in 1984 he exited the lab with a surveillance drone, code-named Amber.

Karem's wasn't the first drone in the United States, but it was the first one to actually work. Each of his devices cost $350,000, which was about the amount needed to run the Army's crash-prone Aquila drone for a single hour.[1] Nevertheless, the project hit a wall of federal bureaucracy, the company was sold to San Diego-based General Atomic, and the Amber was mothballed. Karem persisted, developing for General Atomics a less-sophisticated version for export. He called it the Gnat 750, but no customers stepped forward. By the late 1980s, there was every indication that Karem's work would come to naught. The Amber was killed off in a round of post-Cold War budget cuts. The Gnat 750 met the same fate.

When I met with Karem in his office, he told me what happened next: "It was during the Siege of Sarajevo. President Clinton was frustrated that so little was known about what the Serbs were doing on the ground," including possible violations of the Geneva Conventions. American intelligence agencies turned to satellites in an attempt to document Serb war crimes. While satellite data proved important for military operations, they were blind at night.[2] Breaking the siege required better data about Serb activities, yet persistent cloud cover blocked satellite photos. Complicating matters further, Serbs knew when satellites passed overhead and used this knowledge to evade detection. Surveillance planes like the U2 were vulnerable to being shot down, and they couldn't see through the clouds anyway. To make matters worse, Serb troops strategically dug mass graves at night, thereby eluding satellite surveillance. But CIA Director James Woolsey had heard about the Gnat. He had the unmanned aerial vehicles (UAVs) dusted off from storage in the Mojave, fitted with high-magnification and night-vision cameras, and sent them to the Balkans. By early 1994, two of the awkward-looking devices were being flown out of a clandestine base in Albania.

They weren't perfect. Signal noise from power lines distorted the feed, and control was difficult over great distances. Nevertheless, the Gnat impressed everyone on the project. The impact, Karem reflected, was two-fold: "It proved the value of the technology, and I believe the Bosnian conflict ended four to six times faster because the Gnat was deployed." What's more, Karem pointed out, "the Gnat didn't have missiles! Instead, it had sensors hovering over these [Serb] guys." This proof of concept appears to have been enough for the US military. The platform was refined, armed, and renamed the Predator. Its emergence would profoundly change the nature of armed conflict. Perhaps Paul Virilio was right to argue "history progresses at the speed of its weapon systems."[3]

Talk to Karem for long enough, and it becomes clear that his worldview has been shaped by the Second World War, and the Holocaust in particular. "Look," he told me as we sat in his conference room, the door open to a clear view of the Gnat on display in the hallway: "As a Jew, I'm supposed to be dead, but I'm not because of military intervention by the United States." Karem is oft quoted as stating that he was never the one who armed the drone,[4] something he is at pains to put into context, emphasizing the importance to the United States of both crucial intelligence and credible deterrence.

Nevertheless, in this origin story we can outline an alternate history of the Predator, a device that made its first appearance as a Gnat in what we might now consider to be a humanitarian or "responsibility to protect" role. Critics will find much to carp at in this analogy, but this historical detail suggests a different path the technology might have taken had it continued to be used as a tool to document human rights violations and large-scale crimes against humanity, rather than as a weapon to assassinate "enemy combatants" in a "war on terror." This approach is at odds with a significant literature on the perils of weaponized drones, which overemphasizes the devices' kinetic capacity while perhaps underemphasizing the ethics they create and sustain.[5]

Such an alternate history also highlights the seldom-acknowledged role of satellites in human rights and advocacy work. The Amber was used to complement American intelligence efforts to monitor a rights-violating state. Both technologies hold promise and peril. Here, too, the Gnat/Predator's history is illustrative. In our conversation, Karem made a point of observing that he never armed the device, and even Navy Captain Allan Rutherford cringed upon hearing the newly modified device had been named Predator, since it was "just a surveillance and reconnaissance drone, an eye in the sky. *Predator*, on the other hand, sounded like a weapon. Nobody had suggested arming the new drone."[6]

Once weaponized, Karem's invention was catapulted forward by a decade and a half of battlefield deployment, starting with the United States' invasion and occupation of Afghanistan in 2001, then Iraq in 2003. A desire for victory without sacrifice led subsequent administrations to deploy drones at ever-increasing rates. Reports of drone strikes, both successful and otherwise, dominated media coverage of the wars. Writing in *The Guardian*, North Waziristan resident Rafiq ur Rehman describes his experience of this decision:

> Nobody has ever told me why my mother was targeted that day. The media reported that the attack was on a car, but there is no road alongside my mother's house. Several reported the attack was on a house. But the missiles hit a nearby field, not a house. All reported that five militants were killed. Only one person was killed—a 67-year-old grandmother of nine.[7]

Coverage of this sort has suppressed enthusiasm for drone strikes within the general public and has given rise to a movement against "killer drones" among organizations like the Open Society Foundation, Human Rights Watch, and Amnesty International.

Their argument is simple: subject the United States' drone program to the same levels of scrutiny applied to other war-making efforts. Doing so would require taking drones out of the unaccountable hands of the CIA and placing the program into the military's eco-system, thereby extending existing rules of engagement to this new technology and increasing accountability in cases of misuse. Activists organized die-ins, organizations launched petitions, and artists engaged the topic in installations meant to challenge the West's assessment of events half a world away.

A two-hour drive down Interstate 5 South from Karem's shop will take you to the southwestern-most edge of the United States. Here you will find the city of Chula Vista, which is home to a startup called ActionDrone. The founder was happy to show me around the space, packed with prototypes and custom models mid-production. Devices built by manufacturers like ActionDrone are upending this earlier conceptualization of drones as dedicated killers. Drones may be engaged in "precision" assassinations around the world, but their smaller cousins rely on similar communication and control systems to perform more benign activities. ActionDrone is using a proprietary platform to inspect wind farms for industrial giant Siemens, and has their eye on railroad inspection as well.

It has taken time for this technology to migrate, as the Internet did two decades earlier, from the military lab to the private sector. It took a decade for stable control systems to make their way into consumer-grade technology and to begin showing up on the streets (or sky). Affordable global positioning systems (GPSs) and stabilizers, sophisticated flight-control algorithms, and longer batteries made the jump from other industries at about the same time. The result of this fevered round of innovation can be seen in the rapid sales of small devices available from manufacturers like Chinese DJI, American 3DR, and French Parrot, as well as industrial applications from companies like ActionDrone. At the University of San Diego, a collaboration with engineering colleagues Gordon Hoople and Beth Reddy (now at the Colorado School of Mines) generated classes that bring engineering and peace studies students into the same classroom in order to debate the use of drones, and then build them in heterogeneous teams focused on uses that make the world a better place.[8]

We now build and fly piles of these things with an eye toward the public good.

What are all these devices—drones especially—doing? The clear answer from the human rights world is that they threatening civil liberties, violating privacy, and disrupting terrestrial and traditionally sovereign approaches to security. There seem to be no end of foreboding books, a trend that only accelerated after the 2016 elections in the United States. Yet I hope to write here about a quite different range of uses, and am unabashedly rooting for new tools that have a clear benefit to the public. While this particular round of innovation is extremely dynamic, it is obvious that drones are being used to support human rights, humanitarian efforts, and advocacy uses worldwide. This book is not about where Karem's invention ended up—as a beta test for killer robots—but about how it began: as a Gnat supporting humanitarian intervention. Perhaps a fresh genealogy of the drone can trace a new line from Karem's animating vision to the humanitarian efforts of the future. Indeed, as I write this, General Atomics has announced that its Predator C will be available for humanitarian payloads. Drones, satellites, balloons, and kites are tools in use, and at various times, and for various reasons, have formed particular repertoires. This chapter, and those that follow, trace the opportunities and implications of incorporating these tools into the repertoires of those trying to make the world a better place.

GEOSPATIAL AFFORDANCES

While this book focuses on small drones, the origin of the Predator holds several important lessons. First, it suggests an alternate history of the drone, not as a killer robot, but as a supporting actor in an effort to end ethnic cleansing. The second point is that Karem's early devices were deployed to complement, rather than replace, satellite imagery and a host of terrestrial tools and technologies. ActionDrone, like all other contemporary drone manufacturers, relies on global positioning imagery to increase the stability of their platforms and ensure ease of flight. These system effects are indicative of the ways drones rely on earlier innovations, including the explosive growth of mobile phones, as well as innovation in control solutions from other sectors.

The previous chapter suggested that technology should be thought of as *tools in use* and that affordances are the *possibilities that things offer for action.* In this chapter, I will suggest that an important new range of tools are being put to use in the air, and that these tools cluster into stable patterns of use that we can think of as *geospatial affordances.*

By geospatial affordance, I simply mean those possibilities that *mobile* things offer for action *from the air.*[9] A focus on mobility rules out technology like the closed-circuit television, while a focus on the aerial rules out terrestrial, subterrestrial, submarine, and subcutaneous robots. Of all possible geospatial affordances, I consider but a handful: drones are discussed rather extensively, while satellites, balloons, and kites are discussed in far less detail. Ignored altogether are expensive technologies that already have pride of place in the scholarly and popular imagination: helicopters and airplanes.[10]

Each of the four technologies I reference here—drones, satellites, kites, and balloons—open new spaces for political contestation. This fact should be of some interest to anyone focused on the impact communication technologies have on the emergence and spread of new ideas. My use of the term geospatial affordance draws extensively on the work of Steven Livingston, who has suggested the term applies to the flexible deployment of "spatial and pan-optical awareness and virtual presence."[11] Livingston's work on the use of satellites by human rights advocates leads him to note they make visible "denied access areas using tools that provide verification of eyewitness testimony when available; information even when eyewitness and survivor testimony is unavailable by other means, and types of data that are unavailable to other nontechnical means."[12] He is right, and satellites are not alone in this space.

Geospatial affordances lower the cost of acquiring crucial information about things happening beyond unassisted human sight. Satellites, for example, gather spatial, spectral, and temporal data about the earth.[13] Spatial resolution, most easily thought of as the amount of detail in an image, has grown more powerful as imaging technology increases in sophistication. Yet the quality of the image might not be the most important factor at play in social change efforts. If an environmental advocacy group is focused on rates of deforestation, then spectral resolution might be most important. Spectral resolution refers to the kind of light the satellite's sensors pick up (e.g., ultraviolet, thermal, visible light). Temporal data indicates the frequency with which images are made of a particular area. In quickly unfolding human rights events, having frequent updates in low-resolution might be more important than waiting longer for higher-resolution images. Livingston's work also emphasizes the importance of "temporal reach-back capabilities" of imagery with high temporal resolution. Having extensive imagery of a single place increases our ability to track changes over time. While these factors are most frequently associated with satellite-based earth-observation practices, they apply to balloons and drones as well.

Geospatial affordances increase the data that can be gathered about the earth, but they also increase the tools change-oriented actors have available for action. Livingston has argued that new geospatial, forensic, and networking affordances democratize the process of interpreting what all this data means. Students of social movements call this process *framing*, as advocates set out to define issues, tell causal stories about those issues, and mobilize public sentiment around particular sets of solutions.[14] Geospatial affordances help tell old stories from new perspectives, while also allowing for the emergence of new stories altogether, as temporal data helps advocates to see previously undetectable environmental changes, for example, or when increased spatial resolution brings previously invisible issues into focus for the first time. Lower-cost devices also democratize the process of gathering data that had once been the purview of university and government laboratories. David Hess and his colleagues have documented the challenges involved in having issues of public concern categorized as worthy of official attention—exploring lay efforts to gather data about empirically accessible issues that are overlooked by those with resources and prestige. Citizen science efforts put key resources into the hands of community members themselves, allowing them to challenge dominant narratives of environmental health.[15]

Thinking in terms of geospatial affordances also highlights the importance of spatiality to politics. Drones, for example, are able to maneuver with more freedom than previous technologies. Geospatial affordances reshape the view from the top down, but also produce a kind of *surround sight*.[16] While geospatial affordances can be explored using existing conceptual tools—research and development, the role of regulators, diffusion of technology, reception by publics, embeddedness in sociotechnical systems, and so forth—they also open spaces for and raise new questions about contestation, meaning making, and resistance. In particular, these tools require fresh theorizing of the verticalization and colonization of the ground, the sky, and the subterranean, something discussed at greater length at this chapter's end. New questions about what space is public and which is private will take time for regulators and societies to sort out. New policies and new norms are needed, but will take time to emerge.

But enough with the *geospatial affordance* jargon. What, exactly, are we talking about?

KITES

"Cling, cling, like the lizard, to the ceiling.
Stick, stick close to the side of heaven."

—MAORI KITE-SONG[17]

Let us focus first on the thing that makes me the happiest. Kites. I will tell two histories of the kite. The first is personal. On our first anniversary, my wife and I purchased a kite, agreeing that whenever a disagreement got so strong that it seemed unresolvable, we would go out to fly a kite. Nothing, we decided, would be as soothing as standing there, a string in our hands, staring into the heavens. We've needed it fewer times than we anticipated, and now fly it with our children. Every one of the steps is lovely—the running to catch the wind; the unraveling and unspooling; the sitting, standing, and staring; the handing to a neighbor, friend, or stranger so they can have a go; and then the re-ravelling and re-spooling and coiling as it is drawn back to earth for the folding and packing for storage.

The whole thing is a ritual, a blessing, a sacrament.

I would like to think this has always been the way of it.

Kites are, without a doubt, the oldest of the four technologies considered here.[18] While satellites date back to the 1950s, crewed flight to the turn of the last century, and balloons to the late eighteenth century, the kite's timeline is on a different order of magnitude.

The kite's history disappears into mist, the trail lost in the clouds of Chinese history some two millennia back. Was it perhaps a farmer whose hat was clipped to his shirt, and who noticed it caught in a gale, suspended over his head—aloft and dancing?[19] Or was it the philosophers Mozi and Lu Ban, following after Confucius in the Warring States Period (475–221 BCE), who designed and flew the first kites for some reason that has been lost to history?

The story I like least, but is perhaps the most plausible, involves a siege. During the Han Dynasty (206 BCE–220 CE), General Han Hsin made clever use of a kite to fly over an enemy's battlements, and thereby estimated the distance needed to tunnel under the walls to which his army was laying siege. The estimate was secured, the tunnel was dug, the defending warriors were surprised, and the city was sacked. Perhaps leaning too heavily on magical realism, an alternate version of this story suggests that, in fact, Han was quite small of stature, and it was he himself who was lofted by the kite so that he could espy the enemy's position with his own eyes. The defending forces took him for an apparition, and he obliged, telling them that they should return home or face certain and terrible death. The soldiers fled. In both stories the city was sacked, and its fall marks the beginning of the Western Han Dynasty. This is the second history of the kite.

From China, kite technology diffused along the local trading routes of the day, and during the Han Dynasty probably reached India, Japan, and Korea. Legend has it that kites were used as a brilliant military tactic and effective piece of propaganda during the Silla Dynasty (595–673 CE). The general Gim Yu-sin was charged with putting down a revolt but was unable to mobilize his troops, who, believing a large shooting star to be a bad omen, refused to fight. Gin Yu-sin set out to undo the apparent curse by reversing the arc of the falling star. He used a kite to loft a ball of fire into the air, restoring the star to its firmament and reassuring the men that their efforts to rout the rebels were on spiritually firm footing. The campaign was a success.[20]

Where I suggest an alternate history of the drone emerges from my conversation with Abe Karem, the history of the kite—from trade routes to routing rebels—leaves room to again imagine a number of origin stories. Whatever the case, each of these stories disappears into the mist of love, art, and war.

The kite's path forward is a bit clearer, as the innovation spread from China to its neighbors, and soon well beyond, over trade routes like those developed by Marco Polo, who may have first introduced the device to European society upon his return in 1293.[21] From this eclectic and rumored range of uses a handful of stable applications emerged in military, science, and cultural spaces.

Science—Leonardo da Vinci, who history helpfully reminds us did everything before anyone else, designed a kite-based system for spanning a valley. To the relief of subsequent generations, he didn't get around to building all of the stuff he designed. That task fell, it seems in part, to Charles Ellet, Jr, who in 1848 deployed a kite-based technique for spanning Niagara Falls so that the first suspension bridge could arc the 240-meter-wide gorge.[22]

Ellet wasn't alone, as the eighteenth and nineteenth centuries were a heyday for scientific exploration, and kites were no exception. Inventors lofted science instruments, tested wing designs (biplanes are just motorized box kites), and, perhaps most famously—at least in the United States—were used by the American inventor Benjamin Franklin and French inventor Jacques de Romasto to prove that lightning is made of electricity.[23]

Plaudits for the first scientific use of a kite go to Alexander Wilson, who in 1749—four years before Franklin—used a kite with stacked wings to simultaneously measure temperature at multiple altitudes.[24] For the next century and a half kites served in many roles, continuing to hoist meteorological instruments, inspire or prototype airplane designs (Alexander Graham Bell and the Wright brothers both appear to have tested kites capable of lifting humans), and suspending communication antennas for transmitting multiple frequencies.

But the story I like best is that of the Nobel Prize–winning pioneer of long-distance radio transmission, Guglielmo Marconi, who in 1901 lofted a 500-foot antenna into the air. The goal on a chilly December 12th was to catch a signal sent from a transmission station in Cornwall, on England's West Coast. The antenna was suspended from a Levitor kite flying high over the small town of St. John's in Newfoundland.[25] St. John's was doubtless chosen for its convenient location, and Marconi's team was not the first to notice the location's merit. The team's successful trans-Atlantic wireless transmission took place near the same spot that the first transatlantic submarine telegraph cable made landfall in 1888, and adjacent to the place John Cabot's expedition first made landfall more than four centuries earlier.[26] St. John's then, is an unlikely but important node in the curious stitching of the circuits of capital and empire. The Levitor kite—to return to the focal point in this small story, as Marconi has already made his own mark on history and I leave St. John's to a more able scholar—was the brainchild of B. F. S. Baden-Powell, whose patent for a "man-lifting" kite was granted by the British patents office in 1895.

Scholars of the history of science suggest it was the intersection of the telegraph and qualitative meteorology that produced the idea that climate might be predictable on a global scale.[27] This realization, and a series of international conferences on meteorology starting in 1853, highlighted the need for sensors that could remain stable at particular altitudes. Both the balloon and the kite fit the bill, though the latter was more affordable.

Kites and balloons were considered, more generally, to be platforms for a number of payloads. The American journalist William Eddy used kite trains (stacked kites originally used by meteorologists) to loft cameras to altitudes of 1,200–1,800 meters.[28] Since kites are essentially wings, they were also used by aviation enthusiasts to test a range of theories related to aero-dynamics, though by the 1930s the airplane did most of these tasks more easily. The history of the kite is a tangle of entrepreneurs working across platforms and for any number of motives.

Desire for flight is a dream with many manifestations.[29]

Military—The Cody War-Kite, named after the Wild West Showman Samuel Franklin Cody, was patented in 1901 as a "man-lifting kite" intended to, well, lift a person into the air, thereby giving the military an on-demand vantage point for surveying the enemy.[30] Here, too, I would like to stage a minor intervention: the use of the word "man" in the term "man-lifting" is wrong on two fronts. The first is grammatical and familiar to most readers: the use of the pronoun "man" for "persons" is silly and sexist. But the term is wrong-footed as well as wrong-headed: the first person to be carried aloft by a kite was a woman. Her name, too predictably, seems lost to history, but of the 1827 encounter George Pocock, the inventor of the apparatus, wrote: "We must not omit to observe, that the first person who soared aloft in the air, by this invention, was a lady, whose courage would not be denied the test of its strength."[31]

A spate of experimentation in the late nineteenth and early twentieth centuries was undertaken by the French, Italians, Russians, British, and Americans.[32] These platforms were used to surveil enemy positions, drop bombs, and serve as barrage kites. Legend has it that in Cody's Wild West Showman days, his cook, originally from China, introduced him to the technology. While we have no way of verifying this origin story, Cody indeed went on to develop the device for the British War Office in 1901, used it in the Second Boer War, flew it to an altitude of 2,000 feet over London, and traversed the English Chanel in a collapsible lifeboat pulled by kite. He also experimented with balloons and airships before moving on to the airplane.

There are more stories to be told, but the short career of S. F. Cody (spanning the turn of the century to his untimely death in 1913) neatly captures the rapidity with which kites and balloons were tested, only to be discarded once airplane technology was sufficiently developed. Kites have been of little note in military use ever since, with the very recent exception of their use to firebomb Israel from launch positions in the Gaza Strip.[33] The Israeli Defense Forces responded by guiding a wave of small drones into the kites' strings, ensnaring them in an attempt to bring them to earth.[34]

Propaganda, like all political communication perhaps, lies in the space between military maneuvers, society, and politics. The first recorded use of kites for propagandistic purposes was in thirteenth century China, when a besieged city used kites to distribute pamphlets that incited their imprisoned comrades to escape.[35] During the American Civil War, the same technique was used by Union forces, who fitted kites with pamphlets announcing Lincoln's Amnesty Proclamation to the rebel forces.[36]

Politics and society—Few scholars have explored the kite's social and political implications. My hunch is that there is something of import in the string linking earth and sky, and something significant about individuals performing this act together. I want to say that the kite creates publics, as with the kite fighting—made famous by the novel *The Kite Runner*—that has a long history in Afghanistan and Pakistan.

Events like kite fighting create opportunities for civic engagement, even in the form of friendly competition.[37] In this way, an overarching public sphere is created from scratch. Individuals collectively and intentionally occupy space in the air, on the land, and conceptually as a public. Everyone is doing something altogether, and something altogether has a public and shared meaning: these shared meanings and collective actions take place when everyone deploys creative objects that occupy *and even create* public space. This possibility should give us pause, and it suggests the public-creating effect of kite fighting deserves additional attention from scholars of politics. Why else, we should ask ourselves, would the Taliban prohibit kite fighting in Afghanistan, as the Germans did in France and the British did over England during the Second World War?[38]

I have no desire to carve out new political space for the kite. In fact, I would prefer it remained in the world of art and leisure, discarded by those seeking wealth and power and instead taken into the hands of children and sky-gazers. The Chinese, after all, believed kite flying was good for your health, and that the simple act of releasing the string would turn one's fortune and settle the mind. In Polynesia, some scholars suggest, the kite may have functioned symbolically as a life token, or *liberated external soul*. This is conceptual play, but even here, in perhaps the toughest test case of our four, we encounter the ways in which objects in new space can, for reasons unintelligible in other contexts and perhaps impossible to replicate in other times, threaten the status quo and the powerful.

Artifacts have politics, even the humble-glorious kite.

Kites may have started with warfare and spread through the early but extensive global circuits of private capital, but they have now ended up in the public's hands. Of the four technologies discussed in this chapter, the kite owes the least to Western science, and has turned out to have few features useful for either warfare or commerce. It is simultaneously the oldest, most widely distributed, and most affordable of the four technologies discussed in this chapter.

BALLOONS

Where kite are ancient, balloons are relatively young.[39] Starting in the late eighteenth century, French inventors began consistently experimenting with lighter-than-air flight, and this early lead paid off in a number of breakthroughs.[40] The years 1782–1783 proved to be particularly fruitful, as brothers Joseph-Michel and Jacques-Étienne Montgolfier developed and tested a load-bearing balloon to such an extent that they were prepared, within a few months of experimentation, to host an exhibition flight. Less than a year later, they launched the world's first passenger flight, comprised of a sheep, a duck, and a rooster.

The first human animals to go up in the brothers' device was Jacques-Étienne, and the second was Jean-François Pilâtre de Rozier, who went up later that same October day in 1783. These first flights were tethered and brief. Within a month, de Rozier, joined by a colleague, conducted the world's first free flight by humans—ascending to 500 feet and, in the course of their 20-minute flight, traveling about five and a half miles. Within months of this voyage, the first crewed hydrogen balloon flight took place from the Jardin des Tuileries in Paris. By 1785, the first successful balloon flight across the English Channel was completed by Jean-Pierre Blanchard and John Jeffries, French and American balloonists, respectively. By the 1790s, Americans had begun experimenting with the technology as well. The French philosopher Paul Virilio mused that inside each invention is its accident.[41] This was no more true than in the case of Pilâtre de Rozier. Often considered to be the first person to fly in a balloon, he also appears to be the first to be killed in a balloon. Two short years after his first record-breaking flight, he perished along with his co-pilot, the Marquis d'Arlandes, in an unsuccessful attempt to cross the English Channel.

Originally built and flown by individual inventors for the purpose of adventure and entertainment, balloons were soon adopted by military and scientific communities. The nineteenth century saw ballooning evolve from an area of experimentation by individual inventors to an industry of interest to the nation-state. Their use as observation decks and later as payload-delivery systems (i.e., bombers) proved to be short-lived, as they were difficult to control but easy to spot and shoot down. By the late nineteenth century, two distinct areas of inquiry emerged: the use of lighter-than-air devices to travel and deliver payloads (alternately called airships, zeppelins, or dirigibles) and other balloons (used for entertainment, solo travel, or observation).

Airships emerged in the 1860s as hobbyists, inventors, and entrepreneurs experimented with hydrogen and silk. Military uses were pursued between the Civil War (where they were used to observe the enemy and direct artillery fire) and the First World War, at which point they were written off as a wartime platform. Airship manufacturers turned their attention to peacetime use and developed the dirigible as a form of transportation. The Empire State Building, for example, was constructed with a tall mast for docking transport dirigibles. The British built masts in England, Egypt, Afghanistan, and Canada in anticipation of an empire-wide mail system connected by airships.[42] This plan built off Germany's successful deployment of airships in international transportation. Through the 1930s, hundreds of dirigible flights were conducted between Germany and the United States and between Germany and Brazil. The 1937 *Hindenburg* crash that brought the era to an end occurred within a day of its sister ship's landing in Rio De Janeiro.

Both the *Hindenburg* and the *Graff* were part of a larger aerial effort by the Nazi Reich. In 1936, they were used to drop political fliers from the air and to play patriotic music, political slogans, and political speeches through loudspeakers. The crash of the *Hindenburg* brought this to an end once and almost for all. A brief exception to this airship winter bears noting, and it occurred when the United States launched, in the 1950s and 1960s, high-altitude reconnaissance balloons with the intention of overflying the Soviet Union and China.[43] These flights came to an end when it became clear that satellites and faster aircraft could do the job more reliably. Thus discarded by military elites, the creation, care, and preservation of balloons reverted back to the inventors, adventurers, and scientists who created them in the first place.

Today a handful of these platforms are in operation, with some being used for advertising products and filming events (e.g., the Goodyear blimp), while others are tested by the military-industrial complex in the hope that the technology can finally be refined for transporting heavy payloads over long distances to inhospitable areas. Heavy-lift balloons—imagine sending drilling equipment to the North Pole or tanks to Kandahar—are plagued by the same kind of control challenges that bedeviled their predecessors a century earlier.

For almost a decade, Greenpeace has owned and operated an airship, the A. E. Bates, which they use in much the same way as the Goodyear blimp. In a 2014 campaign to highlight the prevalence and power of the US National Security Agency's (NSA) domestic spying activities, Greenpeace and the Electronic Frontier Foundation flew the A. E. Bates over a large NSA data center in Utah. Greenpeace owns the piloted airship, and previously owned a hot air balloon. These have proven to be high-profile messaging platforms and have helped draw attention to other social, environmental, and political issues, including a Southern California retreat organized by the conservative Koch brothers. However, these appear to be the only contemporary uses for balloons for advocacy purposes.

The dream of long-haul and heavy-load transportation via airship continues to elude investors, and current use is limited to novelty flights. Much the same can be said for a series of solo-ballooning initiatives, as with current efforts to break records related to altitude and duration. Balloons for scientific study have a parallel but less dramatic history. In a pre-satellite era, they presented the most affordable platform for high-altitude tests. The 1886 development of a weather balloon by French meteorologist Léon Teisserenc de Bort represented one of the earliest such efforts, and eventually led to the discovery of the tropopause and stratosphere, the existence of which were unknown to scientists of the day.

While balloons are not in widespread and consistent use by any set of actors this study has been able to identify, they have captured the attention of activists, sometimes to great effect. When the Deepwater Horizon oil rig ruptured, BP attempted to restrict access to the affected area,[44] perhaps hoping to prevent images of the 200 million gallons of crude oil disgorged into the Gulf of Mexico. This embargo prevented the press from capturing images from the air.[45]

Jeffrey Warren, an MIT grad student at the time, used grassroots mapping techniques to help fisherfolk and other community members document the spill's impact. Their imagery demonstrated the spill's scope in ways the embargoed press couldn't. Their footage was made using readily available balloon technology and homemade platforms for their cameras. Warren's effort applied lessons he had learned using balloons to support landless laborers' claims to land in South America.[46] He later pointed out that "there was no publicly available, ortho-rectified imagery available in the initial weeks of the spill," as the public had to make do with lower-resolution imagery from NASA's Terra and Aqua satellites.[47] These grassroots mapping efforts were subsequently funded by the Knight Foundation and eventually became the non-profit advocacy group Public Lab.[48]

As mentioned previously, Public Lab now sells a wide range of do-it-yourself science kits for activists doing what the sociologist David Hess has coined as *undone science*: the rigorous inquiry into problems that have been written out of official scientific discourse and inquiry.[49] For the DIY user, Public Lab sells a modified version of Warren's setup. Balloons and kites share a notable feature—accountability. Want to know who is overflying your home or community? Simply follow the string down to find the terrestrially bound person holding it in their hands, and you have your answer.

Here it bears mentioning that the earliest ballooning efforts, going all the way back to the Montgolfier brothers, were undertaken not by industrialists or governments, but by individual inventors eager to see what could be done. In the American Civil War, the Union Army deployed them reluctantly, and indeed they proved to be an unwieldy observation platform.[50] In the

First World War, the French, Italians, and Germans deployed airships with some enthusiasm. However, it became clear that, once in the air, they were vulnerable as aircraft and unreliable as missile-delivery systems.[51] For all intents and purposes, these debacles ended most governments' interest in balloons as a reliable and scalable weapon in modern warfare.[52] This detour was dark, but relatively short-lived. In the 235 years since the Montgolfier brothers set to work, only 60 were spent in ambitious exploration of their military applications.[53] Balloons might not be widely used, indeed there are few cases in this volume, but I do want to be attentive to technologies that are increasingly available to and used by those traditionally left out of earth-observation efforts. It may be that the future of ballooning lies in the hands of large institutional actors like Google, who have experimented with solar-powered and Internet-equipped balloons as Internet service platforms. It is just as likely, however, that the future lies with community-mapping efforts like those championed by Jeffrey Warren and Public Lab.

SATELLITES

From the Soviet Union's launch of Sputnik in 1957 to the present day, the ability to put sensors into orbit has driven space programs in rich and aspirational countries alike.[54] Satellites created the "intellectual space of globalization"[55] and initiated a televisual era that continues into the present, argues media scholar Lisa Parks.[56] Payloads have gotten larger, sensors have gotten more sophisticated, the number of relevant actors has ballooned, and the orbital space available to satellites has grown more crowded. The decades since that first launch have also seen radical changes in the availability of satellite imagery. While satellite technology has a history of its own, the accessibility and usability of this imagery (also called remote-sensing data) can be thought of in three broad phases.[57]

The first is an era of specialist technicians. At its inception, images produced by the United States' remote-sensing efforts were primarily accessible to those in the specialized field of earth observation. This trend continued with the launch of LAND-Sat I in 1972, which made more imagery available, but only to those with the means to gain access and the ability to render raw imagery into photographs amenable to scientific analysis.

The second era began in 1994 with the passage of the Open Skies Act by the Clinton Administration. The act made a wealth of data freely available to anyone who wanted it. This availability was a breakthrough in accessibility, and it accelerated interest in the use of satellite imagery to analyze deforestation and other environmental concerns. The 1980s and 1990s also saw the launch of new platforms by the European Space Agency and later national efforts by China, India, Japan, and Brazil. These new platforms boasted larger and higher-quality sensors for scanning the earth's surface for visible and invisible light. Starting around 2000, commercial satellite operators began building and deploying their own platforms, reducing reliance on systems owned and controlled by various nation-states.

The third era began in 2005 with the launch of Google Earth. The prior proliferation of platforms meant a wealth of data was available, but its flow to the public has been limited. One factor was price—most simply could not afford to buy imagery from large commercial enterprises. A second factor was the legibility of the images—in many cases imagery needed to be processed by specialists in order to be useful for the average user. A final factor dampening demand was utility—seeing your home from space is cool, but it was not immediately clear why everyone might want, or how they would use, satellite imagery. Google's large-scale acquisition of the imagery required to build Google Earth and to populate Google Maps removed each of these obstacles simultaneously. The widespread availability of Google's interface meant that a host of applications could be easily integrated into this original ecosystem, making it the clear market leader in public-facing satellite imagery.

Perhaps unsurprisingly, what people have chosen to look at has changed over these three eras.[58] Military uses have predominated since the inception of the US satellite program. National prestige, national security, and national interests are bound up with one another in every country's space efforts. Commercial interests have followed soon thereafter, whether for the production and sale of images or the maintenance of the GPS infrastructure that is pivotal to a growing number of terrestrial technologies. Likewise, scientists working in earth and environmental sciences have been using satellite data for several decades. Early work that benefited the general public focused on crop forecasts, tracking storms, and mapping and planning land use. Over the past two decades, these have been complemented by the use of satellites to monitor deforestation, track climate change, assess agricultural extensification, monitor urbanization, map electricity adoption and consumption, and identify polluting factories.[59]

Gradually, satellites have also been used to document human rights violations, what Andrew Herscher has called *surveillance witnessing*.[60] This chapter started with the CIA's adoption of Abraham Karem's Gnat 750 to augment satellite imagery. The layering of technologies presaged US reliance on interlaced satellite and Predator data in its failed efforts in Iraq and Afghanistan five short years later. In 2003, the US Committee for Human Rights in North Korea secured satellite imagery of prison camps in the Democratic People's Republic of North Korea.[61] Images of the camps were combined with first-person accounts of survivors. The firms Digital Globe and Space Imaging Corporation provided the images to the advocacy group, and reporting on the process suggested that American officials had declined to release similar images, citing national security concerns. This early use of satellite imagery for advocacy purposes may be one of the longest-running, as the original study's author, David Hawk, released a fresh report on the country's continued use of prison camps nearly 15 years later.[62] In North Korea, the horrors of forced labor persisted, yet one thing about the latest report was quite different—the images had been obtained from Google.[63]

As a geospatial affordance, satellites have clear strengths and weaknesses. Satellites are positioned in either fixed geospatial orbit, from which they are able to maintain a continuous view of a fixed position, or they make regular and predictable passes over the earth's surface. The strength of the former is that they provide a near-continuous view of a particular piece of the planet. On the downside, they can see only a particular piece of the planet. The benefit of satellites in geosynchronous orbit is that they are able to provide broader coverage, yet are only able to do so at the particular time they are overflying the surface area in question. These problems can be overcome by launching many satellites, yet this solution raises its own unique challenge of cost. While the cost of putting a satellite into orbit has been driven down by competition, commercial-grade and on-demand imagery remain costly.

Social science scholarship on satellites remains thin on the ground, perhaps because they are out of sight and out of mind, "so firmly beyond the visceral worlds of everyday experience and visibility" in the words of Stephen Graham, a British scholar of cities and urban life.[64] Since we use them for so many things, satellites are now "a key part of the public realms of our planet," whether that be in support of military domination, to coordinate GPS-equipped technologies, or helping advocacy groups track human rights abuses.[65] In other words, satellites are yet another technology of the spatial public sphere, critically comprising what the German philosopher and cultural theorist Peter Sloterdijk has called the *inverted astronomy* of earth observation. For Sloterdijk, satellites are a reversed Copernican revolution, allowing us to discover ourselves as if for the first time, as we digitize our increasingly computational planet.[66]

On the surfaces they cover, satellites create publics and politics. Satellites see things the powerful would prefer to keep hidden. Previously invisible rights violations are now subject to scrutiny. Satellites have been used to document war crimes and state violence in Darfur, Zimbabwe, the Balkans, Syria, Burma, Sri Lanka, Nigeria, and the Democratic Republic of Congo.[67] They have helped to identify social, economic, and political inequalities, including by allowing Palestinian activists to better document the expansion of Israeli settlements and control over land, helping to identify tax cheats in Greece (turns out they are the ones with the swimming pools), and illuminating the extent to which a small elite had captured land in Bahrain, leading to a 2011 uprising in that country.[68] Satellites change what we can see of the ground, and in so doing create a new audience for activists and artists. In an earlier era, land art might have been an offering to the cosmos or the gods, but it is now a challenge to the all-seeing eyes of the state and capital, as seen in chapter 4.

DRONES

While drones are used by corporations and governments, I focus here on the use of these platforms *for the public good*. As a result, I have restricted my analysis to nonviolent and nonmilitary uses. The features that first caught my attention about the kind of technology Tautis and I first used in Budapest were its low cost, its ease of flight, and the extent to which we operated without interference.

The most popular of the new drone technologies, and what we used in our efforts, was the quadcopter. Though the quadcopter design can be traced back to the 1920s and 1930s,[69] it wasn't until important questions about control were solved in the early 2000s that it became commercially successful. These theoretical solutions emerged from the lab at the same time as the evolution in mobile telephony, the prevalence of GPS systems, and as steady progress in power storage made consumer platforms possible. One of the earliest companies, Microdrones, was founded in 2005[70] and was soon followed by others, including Parrott (first drone in 2010)[71] and DJI (founded in 2006,[72] first drone in 2013). The year 2012 saw the rapid increase in reports of drone use of all kinds.

This book's evidentiary substratum is a large empirical effort to assess the nature and breadth of nonviolent drone use.[73] My colleagues and I gathered more than 15,000 publicly available reports on drone use drawn from Lexis Nexis, *Motherboard*, *New America*, UAViators, iRevolutions, and weekly reports from groups like the Center for the Study of the Drone at Bard

University. We coded for *purposeful use*—in other words, *the apparent or presumed goal-oriented behavior around a primary intended action*. People meant to do what they were doing, a contrast with a host of reports of flyaway drones and high-profile crashes. We further narrowed our focus on *nonviolent* drone use, thus excluding reports on violent military drone use, especially in high-conflict regions where the United States was actively involved in targeted killing. Manual coding techniques were used to narrow these 15,000 reports down to 1,131 unique, purposeful, and nonviolent drone uses in the six-year period between 2009 and 2015.

The reader may be additionally comforted to know that we controlled for additional terms on a year-by-year basis. For example, "the drone of the vuvuzela" was a popular reference during the 2010 World Cup games in South Africa. Likewise, honey-bee colony collapse was an area that focused on an altogether different drone. The band *The Drones* experienced a surge of attention in 2010, and we wish them well—we dropped them from our sample frame nevertheless.

This exercise shed light on the nature and range of adoption by advocacy groups and change agents and introduced us to many innovative efforts. A sustained examination of these innovative uses generated several of the case studies found throughout this book, allowing us to purposefully sample within a population of reported use. We found that small drones are being deployed in a host of novel ways and that regulators are struggling to keep pace. Drones challenge current regulatory regimes as cameras move to new places, such as over factory farms or crowds of protesters. Governments are acutely aware of their diminished control over both communication infrastructure and national airspace. National governments are not the only ones facing a disruption of the status quo from UAVs. The number and types of uses and users has grown exponentially—from its roots in the military to a crowd that includes artists, activist groups, academic researchers, and private businesses.

Global trends in types of users—The year 2012 saw a dramatic uptick in experimentation by a wide range of actors, falling into seven broad categories. *Intergovernmental organizations*: transnational organizations that share responsibility equally among many national governments. Examples include the United Nations, treaty bodies, transnational organizations, and scientific institutions. *Governments,* including governing bodies, militaries, and police forces, and government use frequently overlaps with scientific inquiry, especially via university partnerships. *Businesses* are private, for-profit endeavors and they too overlap with other sectors, as when a business conducts research and development on behalf of a government. *Science and academia* includes universities, nonprofits, and research and development. *Civil society groups* are nongovernmental organizations, journalists, religious groups, and other civil society groups. *Named individuals* is a category that captures a range of people flying UAVs in their capacity as private citizens, rather than representing another user category. Finally, a large category, *unknown users*, describes all flights where we could not determine who was flying, or why.

The analysis in this volume centers on civil society's use of drones, basically ignoring government and business users, as well as instances in which the user was unknown. It is important to emphasize the difference between two similar-seeming terms. In our study, we use the term "civil society groups" to describe social movement and nonprofit organizations. In this book, I use the term "civil society" to describe non-state and non-business use of drones. This covers individual users as well as the efforts of science and academia, civil society groups, and intergovernmental organizations. I have opted for ecumenicism, whenever possible.

Global trends in types of uses—Categorizing the way these users deployed drones is perhaps foolhardy. Our efforts to use a coding guide based on prior literature quickly fell by the wayside as a proliferation of uses extended well beyond those documented in my earliest work. I expect many other uses we documented will soon appear similarly archaic. This volume focuses on prosocial uses for drones, ignoring efforts that are primarily economic, like agriculture, commerce, and crime, and also ignoring UAV use within the state's domain, including for surveillance and by security forces.[74] Here too there is overlap as, for example, an agricultural drone is developed by a for-profit company focused on increasing crop yields in the Global South, or when a nonprofit uses technology to help small landholders get more crops from their land.

It should be immediately clear that the focus of this study, and this volume, is quite different from militaries' use of weaponized drones. A wide range of excellent scholarship is readily available on the use of drones in the battlespace.[75] This book instead captures those nimble and non-lethal platforms deployed by the kind of actors listed above. One takeaway from our data is that it is far too early to say where this space is headed, or what is likely to happen in the future. Many of the efforts we documented may prove to be unsustainable, and some of the organizations that I worked with as I began this book have closed shop. I therefore am reluctant to use this data to make predictions about the rate and pace of innovation. I am similarly hesitant to make predictions about legislative trends, though I touch briefly on them in chapter 5.

I have no hesitation, however, in pointing to this rapid proliferation in users and uses as evidence that geospatial affordances are more accessible to civil society actors than at any other point in history. Drones—especially in the easy-to-fly and cheap-to-buy quadcopter category—are positioned to do something that neither satellites nor balloons have been able to do: cheaply and anonymously document social, political, and economic phenomena despite resistance from power brokers and elites—in other words, to be disruptive and emergent. Much of this volume is spent exploring that potential and explaining its impacts.

OBSERVATION LAYERS

Each of these geospatial affordances encourage new ways of seeing space. Each operate at altitudes the others are unlikely to enter easily. Each offers particular vantage points, from the wide-angle possibilities of the satellite to the land-on-a-dime capabilities of small quadcopters. These platforms' features can be thought of in comparative perspective, as seen in figure 2.1.

While this illustration radically simplifies some areas of great complexity, it also suggests areas for future research. In particular, it encourages volumetric thinking. This approach echoes the refreshing work of landscape architect Pierre Bélanger.[76] His essay and installation *Altitudes of Urbanization* (figure 2.2) incorporates several additional technologies, additional species, and subterranean layers into one conceptual space. I have chosen to

Figure 2.1
Observational layers: terrestrial camera, low-altitude drone, kite, balloon, satellite.

define uncrewed aerial vehicles by their aerial mobility and to lump them together with other things that have aerial mobility. This categorization process could have taken another direction, instead categorizing uncrewed aerial systems as a subspecies of robots. An inquiry into the possibilities that robots offer for collective action would lead in additionally fruitful directions, since robots are able to traverse the earth's surface, dig and travel underground, and float on and swim under water, and are increasingly able to enter and inhabit our bodies; these are perhaps *volumetric robotic affordances,* able to operate at multiple altitudes and in diverse spaces—from the body politic to our own bodies.

Seeing from the side rather than from above, Bélanger helps us recognize that the underground, the underwater, and the atmospheric are *often overlapping, intertwined, and entangled.*[77] Taking space seriously requires a recognition of spatial power and the air as a *thick, fuzzy, complex space* through which conflicts flow. Spatial risks, as a result, are *relative, temporal, and interconnected.*[78] This perspective fundamentally challenges the way we tend to

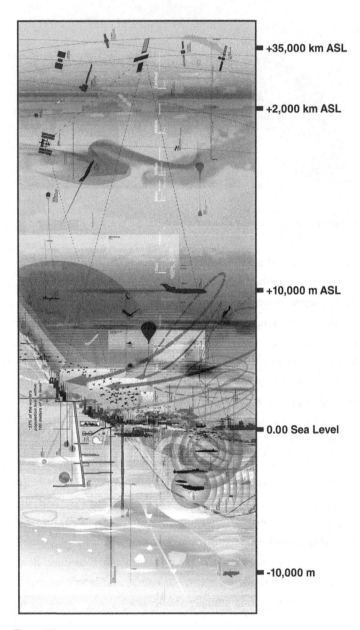

+35,000 km ASL

+2,000 km ASL

+10,000 m ASL

0.00 Sea Level

-10,000 m

Figure 2.2
Pierre Bélanger's altitudes of urbanization.

think about how technology impacts civil society and what kinds of civil societies emerge through our technologies. In a stellar bit of scholarship, Stephen Graham traces this approach back to the ideas found on the cover of Buckminster Fuller's 1923 *Operating Manual for Spaceship Earth*. In the evocative thinker's imagination (figure 2.3), the planet is a node—central, but a node nevertheless—in a network where space and ethics intersect.[79]

For Fuller, the ethical and the spatial are bound up with one another. Thinking spatially is thinking ethically, and, by extension, politically. It is no huge leap to say that thinking politically should also involve thinking spatially. This is obvious to any actor with an army, and as the political anthropologist James Scott has pointed out, *seeing like a state* involves acute attention to space.[80] Identifying natural resources and taxable assets in one's own territory, or seizable land and resources in your neighbor's territory, always involves taking space very seriously. This is as true on the ground as above and below.[81]

FROM CIVIL SOCIETY TO THE CIVIL SPHERE

From their inception in the late eighteenth century through the possible future of autonomous flows of AI-guided aerial delivery systems, geospatial affordances have clear social and political implications: they create new space; their barriers to entry are consistently lowered; and new actors are invited to engage these systems. Such empirical observations have theoretical implications. New actors at work in new spaces necessitate new thinking about the civil sphere. Furthermore, the emergence of new tools in the hands of change-oriented social actors pushes us to think in new ways about how tools are adopted and deployed. In what remains of this chapter, I address both implications in turn.

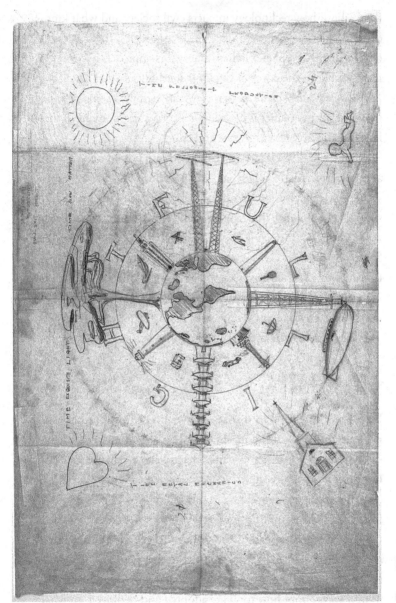

Figure 2.3
Buckminster Fuller's Operating Manual for Spaceship Earth.

Thinking about geospatial affordances begs new questions about civil society and the public sphere. I have focused my attention on nonviolent and prosocial uses of geospatial affiances by civil society and within the public sphere. In the case of drones, especially, geospatial affordances allow the public to hack space—to hack the world of atoms. Civil society is often thought of in one of three ways: as sets of associations like bowling leagues (by Tocquevillians); as a place for discussions about matters of public importance like coffee houses (by Habermasians); or simply as that broad space that is neither state nor market.[82] If I have focused broadly on *change-oriented social actors* up to this point, we can now look at geospatial affordances and ask: *where do they operate?* The short answer is that they operate in spaces falling perhaps into three categories: restricted areas (airports, military bases), private land, and public space. It may be helpful to think in terms of civil society actors operating within a public sphere—and recognize that the public sphere has volume.

What is meant by publics?

We could opt instead for a definition advanced by the critical theorist Nancy Fraser: "What turns a collection of people into fellow members of a public is not shared citizenship, but their co-imbrication in a common sense of structures and/or institutions that affect their lives."[83] This definition is meant to break the notion of the public sphere from its lock-step connection with the nation-state, setting it loose across (horizontal) *transnational* spaces. In the same way, the notion of geospatial affordances pulls the public sphere from its earthly moorings and sets it loose across vertical and *aerial* spaces. Geospatial politics, then, is a conceptual subset of the spatial politics that lie beyond Westphalian horizontal politics and even beyond the politics of verticality.

New work in critical geography focuses on the importance of the vertical in social and political life. Political geographer Stephen Graham is emblematic of this approach: "*As the world's surface becomes more and more congested and contested and urbanisation girds more of our planet, so political and social struggle takes on an increasingly three-dimensional character, reaching both up from and down below ground level.*"[84] Graham tacks in the same direction as theorists and philosophers at work on the politics of space. These include Henri Lefebvre, whose *Production of Space* saw the world as increasingly dominated by the "independence of volumes with respect to the original land,"[85] a theme echoed by the British Israeli architect Eyal Weizman, whose work on Israel/Palestine suggests understanding the conflict is as much about air space as it is about walls.[86] The two work together to articulate power.

Slowly a politics of volume emerges.

A volumetric—or "spherical" in his terms—approach to geographies, imaginations, and public spaces have led the German philosopher and cultural theorist Peter Sloterdijk to write of *air quakes* and other *dangers from the atmosphere,* including gas, as when the air was weaponized as a poison vector by Nazi Germany during the Second World War.[87] The battlefield, and the horizontal terrain of engagement the metaphor implies, is a thing of the past, a vestige from a particular technological era.

The air becomes, suddenly, in the words of Paul Virilio, *battlespace.*[88]

All of this points to questions of spatial politics. This theme is explicit in British geographer Peter Adey's imperative that "we must ask just what kind of life our aerial world has produced as it becomes increasingly the medium for the operation of violence, civil society protest and political power."[89] He goes further: "Just what are the politics of aerial life itself?" This question is echoed and extended by geographers Francisco Klauser and Silvana Pedroso, who specifically argue that drones usher in a politics of vision, and that new tools for engaging the volumetric require new thinking about the politics of visibility, the politics of the ground, and the politics of the air.[90] New conversations about the politics of visibility are refreshingly sophisticated and pick up on a normative nuance introduced by Virilio, who commendably theorizes *vision* machines rather than *surveillance* machines.[91]

Here we find a combination of people and technologies put to all sorts of uses (or, in technojargon, *sociotechnical assemblages with an interplay of actants as they are deployed by a range of actors*). None of this implies routine and systematic surveillance per se, but instead points to the broader and changing regimes of visibility that are at play in contemporary societies.[92] Writers like Klauser and Pedrozo have argued that drones should be thought of as tools for seeing—for vision and visibility—and not just as tools for surveillance. This more inclusive approach recognizes the fact that drones create *unsystematic visibilities* in the hands of public and private users.[93]

These new ways of seeing benefit the powerful and powerless alike.[94] The aerial turn in critical human geography driven by Adey, Elden, Graham, Klauser, Pedrozo, and so many others suggest something more complex than a beta version of *Terminator's* Skynet. They also open opportunities for new practices and new actors to engage in politics in new places, requiring us to explore not only surveillance, but also how new ways of seeing become and create new social and political issues.[95]

The air can no longer be thought of exclusively as a space for hegemonic conquest and ubiquitous surveillance. It is also a site of contentious politics. Increasingly, Graham argues, "the struggles over the right to the city, to living space, to resources, to security, to privacy, to mobility, to food and water, to justice—and even ... the right to live rather than die—are increasingly shaped across vertical as well as horizontal geographies of power."[96]

These struggles are not one-sided.

The public sphere is comprised of those places where matters of public concern are made visible. This includes the sky, especially as it is increasingly occupied by devices. Returning to Nancy Fraser, what makes a public—what turns a collection of people into members of a public—is *co-imbrication in spaces that shape our lives*.[97] This includes the space around us. Publics are emergent properties of proximity, but publics are also temporal and spatial phenomena.

This linking of space and publics is not as radical as it sounds. The built space of architecture has always mattered to the establishment of a public. In France, the great debating spaces of the *salon* and the *précieuses* took their cues from the great hall in a royal court. A diminishment of the Court after the Glorious Revolution shifted attention to towns and their vibrant coffee houses.[98] The democratization of debate was preceded by the democratization of *space for debate*.[99]

A public sphere, in this line of thinking, is something like an emergent property of public spaces where politics and politically relevant actions happen. By emergent property, I mean to suggest that the "public sphere" is not just the process of discursive engagement, but also a thing with volume, a *place for politics*. These include public spaces (coffeehouses and piazzas), cyberspaces (social media and virtual reality), and aero spaces (as argued in this volume). While most scholarship focuses on the discursive and communicative elements of political debate in the first two spaces, a significant amount of politically and policy-relevant activity in science, technology, medicine, education, and art happen in *all* of those spaces.[100]

By arguing for the expansion of the concept of *public sphere,* I am deliberately running together notions of social and physical space. This is an old observation, and once again I lean heavily on others. For Jurgen Habermas, the quintessential public sphere took the form of space *and* process, of architecture *and* debate. The former lays the groundwork for the latter. As already seen, this is also a lesson from critical geography. Emphasizing the intersection of physical space and social meaning pushes us to focus on verticality and volume. Within science and technology studies, Langdon Winner has argued "technologies are structure whose conditions of operation demand the restructuring of their environments."[101] It is not technological determinism to observe that things have effects.

Geospatial affordances require us to think about space, and this requires more critical thought about both verticality and volume in social, economic, and political terms. Important new work by Stephen Graham and Lisa Parks emerges alongside the classic work of Sloterdijk and Virilio, and pushes us to look down from space, up from the ground, and underground as well. But geospatial affordances—especially drones, kites, and balloons that can hover and stare—require volumetric thinking.

If the satellite—invisible to those it watches—exudes control while evading accountability, the balloon does quite the opposite. Low-altitude balloons are often controlled by means of a string. I have a large balloon and a 2,000-foot roll of string on the shelf in my office, and have found occasion to use it from time to time (image 2.4).

The social impact of the string is profound: it allows anyone who can see the balloon to see the person operating the balloon. It is old-school analog accountability at its best. It is the kind of accountability that drones and satellites elude by design. Community mapping via balloon invites community members into the process of image gathering (by holding the string) and image making (by providing their input into the map-making process). The string, I want to write in homage, is a metaphor for technological accountability. It embeds seeing within relationship, context, and perhaps even community.

Here we have the technological antithesis to Donna Haraway's oft-cited *gaze from nowhere*. The view from nowhere is "tied to militarism, capitalism, colonialism, and male suprem-

Figure 2.4
Tautvydas Juskauskas (left) and the author (center) training independent journalists in Central Europe.

acy." The view from nowhere tries to "distance the knowing subject from everybody and everything in the interests of unfettered power."[102] The view from nowhere eschews accountability. An aerial view from *somewhere*, on the other hand, exudes accountability. The view from somewhere links the curious explorer to engaged publics by means of a simple thread. Different technological forms suggest different forms of publicness, accountability, and power. The humble string is evidence of this fact and should be a lesson to us all.

II ITERATIONS

3 HACKING SPACE: NEW TOOLS IN THE AIR CHANGE POLITICS ON THE GROUND

Buzzing high above the moonlit landscape, a six-rotor UAV hovers near a roadway on the edge of South Africa's Kruger National Park.[1] With sophisticated imaging hardware, the device captures the heat signature of an endangered black rhino and beams coordinates back to a command post. An infrared night-vision scan of the surrounding area reveals a vehicle, out of which jump three men who begin scaling the perimeter fence of the park: poachers. Waiting for word from the command post, park rangers are deployed near the location of the rhino, ready to intercept the threat. The word is given, and the rangers move in, arresting the poachers and preventing another rhino from being killed for its horn. As drone use spreads worldwide, efforts like this are on the rise. How are drones being used, and how should we categorize such use?

Across a number of recent real-world cases, this chapter provides empirical support for the concepts introduced in earlier chapters. I began the book with the claim that civil society uses technologies before and beyond social media, and that such technologies include the things people and groups put to use. Do this often enough and a repertoire emerges, virtually always in reference to dominant social, economic, and political resources and norms. These things are put to use when people think things can be put to use. This use can be new (i.e., emergent) and it may challenge the status quo (i.e., disruptive), or it may be both, or neither.

In the second chapter I introduced new tools for seeing from the air ("geospatial affordances") as a particular set of technologies—both old and young—that are independent of social media. I also implied that these technologies represent an emerging repertoire amongst civil society actors. The implications, I suggested, are profound insofar as *spatial* technologies create new puzzles and opportunities around civil society, pushing us to take the concept of the public *sphere* more seriously and more multidimensionally, if you will.

This third chapter is entirely empirical, and allows us to bring the theoretical argument made in the first chapter together with the conceptual argument advanced in the second. In particular, I present evidence that geospatial affordances are tools that change-oriented social actors put to emergent and disruptive use. In the first chapter, I introduced the concepts of emergent and disruptive technologies. There I was unpacking the phrase *disruptive new digital technology*. By now it should be sufficiently clear that by technology I simply mean *things in use*—whether old or new, digital or analog, so long as they help people, groups, or institutions to get things done, preferably for the public good.

The term *disruptive* simply means a technology for which there is little approval for either means or ends. A non-disruptive technology is one that enjoys broad and unproblematic support. One can imagine that the technological repertoire of most nonprofits is decidedly non-disruptive, since nongovernmental organizations (or congressional representatives and churches, for that matter) frequently rely on a range of donors who are themselves imbricated within broad social, cultural, and political norms. The term *emergent* simply refers to whether a particular task can be accomplished with current tools and technology. Of course, when it comes to geospatial affordances, drones like the Predator are not necessarily an emergent technology for well-equipped countries like the United States. The Predator does a few of the things that an F-16 aircraft is capable of, but at a drastically lower price and without risk to a pilot's life. A price advantage does not make a technology disruptive if the actor's funds are nearly unlimited. But for the kind of non-state actors covered in this study, aircraft of any sort tend to be exorbitant, and if they are engaged in politically sensitive activities, or making claims against commercial or state elites, then it is unlikely permission to fly would be granted, even if an aircraft could be procured.

Here we note an interplay between dominant norms and emergent use. In a weak conception of emergent use, I have included politics (and thereby social norms) in my assessment of what is possible. Whether something "can be done" with a helicopter by a civil society group, for example, is directly connected to the actor's budget and official approval to fly in public airspace. In many cases, the phrase *cannot be done by other tools of the day* might more accurately be rendered: *cannot be done by tools of the day because of resistance from both the government (which issues permits to planes and helicopters) and society (which supports organizations with approval and finances)*. The relationships between these factors can be seen in the table 3.1 and is discussed in the case studies that follow.

A number of these cases—anti-poaching (non-emergent and non-disruptive); environmental advocacy (emergent and

Table 3.1 Cases in Book Categorized by Emergence and Disruption

		Emergence	
		Can be done with current tools (non-emergent)	Cannot be done with current tools (emergent)
Disruption	Follows norms (non-disruptive)	*Definition:* Could be done with *other tools of the day* Stakeholders approve (or ambivalent)	*Definition:* Could not be done with *other tools of the day* Stakeholders approve (or ambivalent)
		Example: Anti-poaching (chapter 3) Intimate partner violence campaign (chapter 6)	*Example:* Environmental advocacy (chapter 3) Photojournalism (chapter 6) Anti-slavery lantern shows (chapter 6) Animal rights (chapter 6)
	Challenges norms (disruptive)	*Definition:* Could be done with *other tools of the day* Stakeholders disapprove (or targeted)	*Definition:* Could not be done with *other tools of the day* Stakeholders disapprove (or targeted)
		Example: Crowd estimation (chapter 3) Anti-slavery petition (chapter 6)	*Example:* Documenting war crimes (chapter 3) Slavery in India (chapter 3) Photojournalism (chapter 6) Forced labor in North Korea (chapter 3) Monitoring police (chapter 5) Drone graffiti (chapter 5)

non-disruptive); crowd estimation (non-emergent and disruptive); chronicling war crimes (emergent and disruptive); and documenting bonded labor in India (emergent and disruptive)—are presented in this chapter. Some cases were presented in earlier chapters, and others will be introduced throughout the rest of the volume. In every case, I have done my best to focus the empirical evidence on these two key factors—emergence and disruption. The utility of this approach is tested through the inclusion of examples that rely on technologies other than those covered in this volume: in the final chapter, I explore the implications for a materialist approach to advocacy technology with the case of an intimate partner violence campaign in contemporary South Africa and the use of lantern shows and anti-slavery petitions during the abolitionist movement in England. Many of the cases categorized in table 3.1 are incorporated into chapters 4–6. This chapter focuses exclusively on five case studies that illuminate the utility of the emergent/disruption framework for understanding technological adoption.

ANTI-POACHING ADVOCACY: NON-EMERGENT AND NON-DISRUPTIVE

Thomas Snitch and his team from the University of Maryland's Institute for Advanced Computer Studies have been using a UAV to understand the patterns of poaching that take place in Kruger National Park. The team's data analysis provides a model that directs rangers, using satellite data, to the site where a potential target animal will be at a given time. The UAV provides aerial observation and alerts the rangers when an attempt to kill the animal might take place. Using this complex network of data, technology, and human intervention, Snitch and his team were able to stop poaching entirely within a week of the UAV's introduction.

Although these technologies thwart poachers, legislative challenges threaten their use. Kenya, a poaching hotspot, instituted a broad ban on drones, effectively halting anti-poaching projects. These regulatory measures were far-reaching and have impacted well-intentioned efforts aimed at what seems to be a positive outcome. With a lucrative black market, however, the disruption of these criminal networks can have a consequential influence on corrupt politicians and policy makers, particularly in less-developed countries where economic opportunity often comes from elected positions. While drones may serve as technical tools for problem solvers in civil society, they cannot solve underlying social, political, and economic issues that make advocacy interventions necessary in the first place. Perhaps, then, the use of UAVs in such contexts is more disruptive in countries where networks of corruption facilitate poaching.

Projects like the one in Kruger are becoming more common as drone prices fall. Protecting endangered species with the use of UAVs has created a number of innovative ventures, from the sophisticated tracking team in South Africa to a hacked drone that releases pepper spray to deter elephants from entering an area where they may be in danger. Orca pods are being monitored by a UAV in order to observe the health of their members and to determine whether new offspring are present. An advocacy group called Sea Shepherd is using drones to combat Japanese whaling missions and illegal seal hunting.

Larger organizations are getting involved as well. The World Wildlife Fund has been awarded a $5 million grant from Google for innovative UAV-based approaches to conservation issues, particularly poaching. The project uses a UAV that can fly for an hour and cover a distance of 18 miles at an altitude of 650 feet, thereby expanding the battle against poaching in Africa and Asia by providing information regarding animal locations, danger zones, and ranger deployment. Working together, these efforts have had a significant impact on the ability of poachers to function while also leading to an increase in arrests of potential poachers. In some cases, the areas where UAVs and satellites are used in conjunction have completely eliminated poaching attempts.[2]

None of this is to say the deployment of drones is unproblematic. Recent work by the University of Minnesota focused on how an animal reacts to the presence of a UAV. Researchers strapped heart-monitoring devices onto bears, hoping to monitor any change in heart rate when a UAV was present. Additionally, they used a GPS tracker to see if behavior would be altered by the presence of a UAV. The researchers noticed a significant change in their heart rate during each of their flights, but they did not notice any changes to the bears' behavior. Clearly, there is much more to learn as UAVs join other methodological tools in the conservationist's toolkit.

A casual look at our dataset suggests most nongovernmental and noncommercial uses are neither emergent nor disruptive. In other words, other tools are available for the job (so it is not emergent) and the deployment of drones for the task does not challenge existing cultural or political norms (so neither is it disruptive of the status quo). In the pages that follow, I highlight those cases that represent a scale shift in how individuals and institutions are able to see and act in the world.

DOCUMENTING NONVIOLENT PROTESTS: NON-EMERGENT AND DISRUPTIVE

Drawing attention to the worthiness of claims-making efforts is not new. The first use of a drone to document a protest event appears to have occurred in 2011 at a pro-democracy event organized against Russian president Vladimir Putin. Tens of thousands of people gathered at Bolotnaya Square in Moscow in response to the results of that year's parliamentary elections, widely seen as yet another round in Vladimir Putin's consolidation of power.[3] Ridus News Agency, a Russian citizen journalism group, flew a six-rotor drone mounted with a Canon DSLR camera over the protests, documenting the expanse of the crowds while hovering over a nearby river.[4]

Drones have been a consistent presence ever since. Civil society actors have used them to push back against anti-democratic regimes in places where democracy is under threat, including Hong Kong, Hungary, Turkey, Thailand, China, and Russia. In Ukraine, for example, a quadcopter was used to film anti-government protests in the city's capital, Kiev, after President Viktor Yanukovych announced a shift away from the European Union. The video was later shared online.[5] When an activist used a micro-drone to document political protests in Istanbul in the summer of 2013, the police responded by shooting the device down. However, this was not before the operator was able to capture footage of police-on-protester violence, including the use of water cannons, plastic bullets, and tear gas.[6]

The same year, a Thai drone pilot shot footage of protests in Bangkok,[7] in which supporters of the opposition party called for Prime Minister Yingluck Shinawatra's resignation.[8] The footage of the violent clashes between protestors and authorities—full of angry shouts, water cannons, barricades, and tear gas—directed significant international attention to the protestors' claims. In response, the subsequent junta government banned personal drones equipped with cameras.[9] It is clear why authorities ground drones: they provide footage that is simultaneously a source of inspiration (thus amplifying the movement's effect) and information (providing acute estimates of crowd size, thus further challenging the target's legitimacy). This demonstration effect, if you will, is important for protests in democracies and non-democracies alike. As such, it links to arguments advanced by political scientists Erica Chenoweth and Maria Stephan regarding the relationship between protest size and movement success—larger nonviolent protests are more likely to be successful than smaller and violent protests.[10] Drones are not the only tool for this, but repressive regimes are loath to release to the general public any inspirational and informative photographs of large anti-regime protests.

Together with colleagues at Central European University, and now colleagues at the University of San Diego and University of Nottingham, I have worked to improve on existing estimation approaches for analyzing images of large crowds (figure 3.1).[11] This puzzle has plagued movement scholarship for some time, as efforts to draw estimates from the air are expensive and involve cooperation with the authorities—two things that are nearly impossible to secure for the average social movement intent on disrupting the status quo.

Protests and other large-scale collective-action efforts are often intended to challenge and transform particular patterns of thinking or behavior. Efforts to document these events are similarly disruptive, especially when security personnel are involved. Police are often aware of the risk posed by footage that contradicts dominant narratives about the nature of the protest event, participant behavior, and police response. This is as true for drone footage as it is for street journalism. While documenting protests by drone may be disruptive to established interests, it

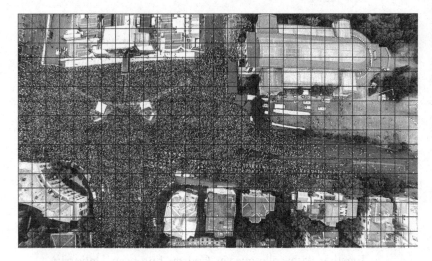

Figure 3.1
Using drone-based digital imagery to better estimate the size of large crowds (author image).

is debatable whether it is emergent. Non-identical, but similar, images can be captured from the top of nearby buildings. Identical images can be made by helicopters granted permission to fly over an event. Here we are reminded of a critical tension—emergent for whom? Drones are an emergent technology when benchmarked to civil society groups, which tend to not have helicopters or access to helicopters, but this same technology is non-emergent to states, which own and control helicopters of their own.

The use of drones to document large-scale political gatherings is disruptive. This is true whether or not their use is emergent. In settled democracies, social change efforts succeed or fail based on their ability to sway elites with policy-changing power, either through direct threats and appeals, or by proxy struggles in the court of public opinion. Over the past 60 years, Charles

Tilly has argued, these efforts have increasingly taken the form of large gatherings and protests intended to demonstrate worthiness, unity, numbers, and commitment.

While capturing the public imagination and changing individual minds is a tall order, an important first step involves demonstrating that a new perspective is held not by a lunatic fringe but by a sizable number of fellow citizens. Numbers matter. Small events are dismissed as fringe, and their issue further marginalized. Visually arresting images of large crowds and unambiguous estimates of crowd size are important for claimants, as anyone familiar with the crowd at Donald Trump's inauguration can readily attest.

Moving images and firm estimates are hard to secure from the ground, but both can be easily captured from above. Social movement actors are rarely in possession of the resources necessary to deploy helicopters, so they have traditionally made do with rough estimates and broad claims. Drones, however, provide an accessible and affordable tool for activists (as well as journalists and the police) to document the event and better substantiate their claims. Of course, the technology might just as easily demonstrate that the event was poorly attended or that some participants broke the rules of engagement. The technology may also be used to increase transparency and accountability, especially in areas where governments, corporations, or other powerful private actors are intent on keeping secrets from citizens and consumers. There is a place for muckraking journalism, to be sure, and the most pro-public acts of journalism in the past century have fallen on the right side of justice but the wrong side of the law.[12]

Of course, advocacy groups and independent journalists are not the only actors at work during protest events. In India, where nongovernmental drone use is banned, the police have made extensive use of the devices for monitoring large political events. When conflict broke out between Muslim and Sikh communities in Saharanpur, Uttar Pradesh, after a land dispute turned violent, local police used drones to monitor the situation. The UAVs were particularly useful in areas police could not access by car or foot.[13] In Lucknow, also in Uttar Pradesh, the city police have used drones for several years to monitor a major religious festival that has lately led to sectarian clashes. In 2015, the police purchased four new drones, now equipped with the ability to release pepper spray against mobs or violent protest.[14] A non-lethal "riotcopter" can be purchased from a South African weapons manufacturer, complete with tear gas, paintball rounds, and a remote speaker system. It is unclear whether these trends will make their way to other countries. Over the past half-decade, an unspecified number of police departments across the United States have begun obtaining UAVs. Only time will tell whether changes in political regime and popular opinion will encourage their deployment.

Finally, drones document protests, but they can also be used as a vehicle *for* protest activities. In an awareness-raising publicity stunt, Dutch activist group Women on Waves used a lightly modified consumer drone to transport abortion-inducing pills across the German-Polish border. The group wanted to raise awareness about restrictive anti-abortion laws in Poland—which offers abortions only in cases of rape or incest, risk to the mother's life, or severe fetus malformation—and to draw attention to discrepancies in abortion access by country and wealth.[15] The delivery went to two Polish women who used the pills. In a similar effort, Japanese activist Yasuo Yamamoto landed a drone on the roof of the Japanese Prime Minister's office in a protest against the use of nuclear power. The drone carried a camera and a small plastic bottle filled with radioactive sand from the site of the 2011 Fukushima nuclear reactor meltdown.[16] No one was harmed, but the incident prompted fears of future attacks being carried out remotely.[17] Such use is non-emergent—the pills could have been smuggled by land and the sand-filled bottle could have perhaps been delivered by trebuchet—but all of the cases in this section are disruptive of the status quo.

DOCUMENTING ALEPPO: EMERGENT AND DISRUPTIVE

In the course of writing this book, some of the world's attention focused on the plight of citizens trapped in Aleppo, Syria's largest city.[18] President Bashar al-Assad faced protests in 2011, but held fast against the kind of anti-regime protests that brought down the leadership of several of Syria's neighbors. By mid-2012 the struggle between the regime and protestors had turned into a civil war.

Over the course of the conflict, Aleppo became more than a city. The brutality of the siege against its civilians reduced the city to rubble and generated countless refugees, leading observers to list it alongside other humanitarian tragedies: "Berlin, 1945; Grozny, 2000; Aleppo, 2016" in the words of the *New York Times*.[19] *Times* journalist Michael Kimmelman's observations were made in response to footage made by a drone as it wove silently through narrow streets, climbed carefully past floor after floor of rubbled homelife, and pulled back for a panoramic shot of the absolute desolation of total war. Such footage represents a significant challenge to the government, especially in light of the military's use of barrel bombs and chemical weapons. Documentary evidence—whether from the air or ground—of war crimes will be vital to any future case against the Assad regime.

In the case of Aleppo, the struggle to frame the conflict for the international community ran parallel to the battle for control over the land itself. In work AlHakam Shaar and I did reporting on reporters, we found that while drone footage from *Russia Today* showed videos of rebel-held and heavily bombed Eastern Aleppo, the regime's Ministry of Tourism promoted footage of the city's intact Western half, complete with the soundtrack from HBO's hit show *Game of Thrones*.[20]

The contrast was stark. Life with the regime was normal. Life with the rebels was hell.

Citizen journalist Monther Etaky remembered that "the regime was always looking down from the drone" and that their footage was used to break the will of the resistance. An Aleppo native working as a journalist during the siege, Etaky and his colleague Abdalrahman Ismail were frustrated by this distortion and desperate to tell a story of resilient defiance. They were not alone. They joined a handful of independent journalists in order to tell the other side of the story. The next step was as simple as it was familiar to activists the world over: they scoured the marketplace of techniques and technologies for the right tool for the job, and then they bought their own drone. Suddenly, the journalists were working along two frontlines simultaneously: one physical and the other symbolic. "The regime was always looking down from the drone. When I first flew the drone for myself, I saw the destruction of the city. I'm used to seeing the destruction from the ground, but not from the sky...it looks wider than from the ground," Monther remembers. At the time, both men were contributing to *Life in Aleppo,* a grassroots effort to raise awareness of the siege. "The regime is the greatest criminal on the planet," Monther told me. With Assad's planes occupying the skies for the past five years, Aleppo earned titles such as "the world's most dangerous city" and Syria's "barrel-bomb capital."

The group's footage undermined official narratives of the war's progress while challenging humanitarian consciousness worldwide. This is true in struggles well beyond Syria, as new technologies give regimes new means of control at the same time that challengers gain new tools for documenting abuses and spreading the word about important causes. Drones are no exception: the most-viewed drone footage of Aleppo is not from Monther, Abdalrahman, or their colleagues. What folks watch the most is YouTube footage from Russian outlets like *Russia Today* and *Ruptly*. Some titles are generic: "Drone Footage Captures Devastation of East Aleppo"; others are clearly political: "Drone Footage Shows Fierce Clashes between Syrian Army & US Backed Islamic Terrorists."[21] That contrast couldn't be clearer, Abdalrahman remembered: "When we are besieged, we all the time see the drones of the Assad regime flying over the city," their footage "telling lies."

Drones allow advocacy groups to see over walls, peer deep into inaccessible rainforests, and capture footage from just across town. Indeed, one of the first things Monther and his young colleagues did was to fly a drone over their university, which they hadn't seen in five years. When they first started flying, people assumed it belonged to the regime, "They said *it's a spy drone, we should shoot that drone down, so it's not targeting our neighborhoods.*" Frequent flights and some neighborly outreach prevented the drone's downing by friendly fire. Nevertheless, Abdalrahman and Monther estimate that they lost 20 drones over the course of the conflict. These losses are due to risky flights that basically involve "gambling to take good footage from regime areas," but they are also the result of the regime's efforts to shoot down their devices or jam their control signals. Such are the basic back-and-forth struggles between regimes and challengers. For now, the aerial struggle has subsided, as both Abdalrahman and Monther fled Aleppo as Syrian and Russian troops moved in. Their departure was marked by a final insult, Monther remembered: "I lost three laptops and a drone—the Russian officer stole it from the bus where I was. As part of the forced evacuation agreement, police were not allowed to open bags—guns weren't allowed, and I didn't have guns, but they saw the laptops, which is the worst gun for them."

Monther and Abdalrahman are citizen journalists as well as activists. They plied their trade with laptops and mobile phones. Did the addition of UAVs to their repertoire matter? In the face of the total destruction of Aleppo, such questions may seem beside the point. But when we think of journalists' work as witness to war crimes, it is important to ask questions about how data has been gathered. Drones do fly in new places—over volcanoes and endangered flora, for example. This usage is hardly disruptive, as few political or economic interests are threatened by affordable data from previously inaccessible bits of the earth. To say there are few examples of emergent and disruptive drone use is not to say that there are none at all.

In conflict zones, UAVs have been used to document the scale of destruction, but they have also been used to document the fighting itself. Flying directly over firefights is what qualifies drone use as emergent in this case. Even prior to the broad availability of affordable UAVs, it was not possible for non-military aircraft to overfly active conflict. The piloted aircraft would have simply been shot down. True, journalists on foot can follow the action. But drones do them one better by gaining a perspective that ground-bound journalists cannot, and by doing so at a proximity that helicopters cannot risk. A bird's eye view of the conflict zone would be impossible, since surface-to-air missiles and groundfire could quickly disable a helicopter or aircraft. The emergent properties of this usage are directly tied to the fact that images can be made without risk to human life (rather than the lower costs of running a drone in comparison to a helicopter). Aleppo fell as I wrote this chapter, but both sides continued to use UAVs. For civil society, drone footage capturing active firefights and their aftermath have the potential to fundamentally challenge official narratives about the nature of tactics and targets.

ENVIRONMENTAL ADVOCACY: EMERGENT AND NON-DISRUPTIVE

Emergent and non-disruptive uses emerge in areas where UAVs are the only way to get the job done, but doing the job doesn't ruffle anyone's feathers. Drone use in scientific research and environmental conservation provides an ideal example. In his work on food shortages in the ocean, Griffith University's Dr. Jan-Olaf Meynecke worried that the health of humpback whales might be compromised in ways unseen from shore. He turned to UAVs for help. Meynecke positions drones over whale's blowholes, allowing him to sample exhaled mucus every few minutes.

The samples are sent for DNA analysis in his lab at Australia's Griffith University.

This type of research was not previously possible. Sampling DNA of live whales living in the wild required the use of boats and crew, which disturbed the whales' behavior. Additionally, boats and crews are expensive and intrusive. Alternative aerial sampling methods, such as the use of large, fuel-powered remote helicopters, are more dangerous, significantly louder, and substantially more disruptive (though researchers attempted this approach in 2009). However, using the DJI Phantom, a popular model he purchased in 2013 for about US$700, Meynecke was able to capture information once out of reach. "The fact that drones have become more affordable and easier to control, with more air time, provides a completely new dimension for research," Meynecke explained in an email conversation with my colleague, Elizabeth Cychosz.

Here in Southern California, the local zoo has branched out into drone research, this time in the Arctic. Collaborating with engineers from Northrop Grumman, a team from the San Diego Zoo Institute for Conservation Research has launched drones in Manitoba, Canada. Using a UAV engineered to operate in Arctic conditions and do so more quietly than a helicopter, future researchers can ultimately observe bears in places where human researchers cannot realistically go. One goal is to research bears in active ice areas, as climate change may affect whether and how bears move from land to ice. "It can give us an unseen eye into polar bear life," researcher Nicholas Pilfold said in an interview with our local NPR affiliate.[22] "We're trying to push the technology to go to regions where humans rarely go and areas that are difficult for us to access."

It is not just biologists who are using UAVs. Scientists in fields as varied as archaeology and meteorology have adopted them as tools for collecting data and exploring new environs. Incorporating drones into scientific research projects has enabled scientists to conduct research activities that would have formerly been unsafe for humans to conduct themselves. For example, a California Institute of Technology astrobiologist teamed up with a drone-equipped filmmaker to explore the lava lake at Marum Crater, Vanuatu. Their goal was to create a 3D map of the lake and sample its soil for life.[23] In another example, archaeologists at the University of Arkansas and University of North Florida used UAVs to collect thermal imagery of archaeological sites to reveal previously unidentified structures, obscured for centuries by erosion and foliage.[24] "[Drones] are on their way to becoming this indispensable and revolutionary technology," Adam Watts told an interviewer at the journal *Nature*.[25]

Scientists run into many of the same legal hurdles as other drone users, with local legislation often limiting where they may conduct certain types of research. Meynecke referenced a three- to four-month process to acquire permits in Australia necessary to carry out his research on whales. However, because many of these projects are funded at least in part by governments, such as with the United States' NASA or National Oceanic and Atmospheric Administration, scientists have an advantage when seeking permission. In mid-2015, for example, the Federal Aviation Authority granted permission to a research collaboration based out of the University of Colorado, Boulder, to fly a drone over a 54,000 square mile area across Texas and Oklahoma in order to study tornadoes and other extreme weather events—a scope not easily imaginable in other contexts.[26] In subsequent years such permits, and permission to fly *beyond line of sight*, have become more common, but it is often major corporations who are best positioned to secure these opportunities.

An ability to work effectively—if slowly and bureaucratically —within the confines of existing laws and policies reflects how this type of use is both emergent and non-disruptive. Using drones for conservation and scientific research enables scientists to study parts of the natural world that were previously inaccessible, marking this use as emergent. It is non-disruptive because scientific research is a widely accepted priority and norm. Although researchers are currently discussing the ethical implications of drone use for scientific and conservation purposes, the use itself is often an extension of traditional and accepted research practices.[27]

SLAVERY FROM SPACE: NON-EMERGENT AND DISRUPTIVE?

Globally, slavery flourishes in unregulated economic sectors and overlooked social spaces. Slavery is illegal worldwide, but is practiced globally. The social scientist and anti-slavery expert Kevin Bales, my colleague at the University of Nottingham, has pioneered research on the topic, and his findings are sobering.[28] Tens of millions of people still live in slavery. Slavery, writ large, is when one person holds another through force, fraud, threats, or coercion for the purpose of economic exploitation. The forms of exploitation vary widely and include trafficking for sexual exploitation (sometimes called sex trafficking), forced marriage, wartime slavery, and bonded labor.[29]

Half of the people living in slavery worldwide are estimated to live in a belt that spans Pakistan, India, and Nepal. I wrote my last book about exploitation in India and spent countless hours trekking through the hinterland to interview perpetrators and survivors of debt bondage.[30] A debt bondage system turns small loans into long-term obligations, and this economic exploitation is amplified by a culture of caste inequality. The result is millions of lives trapped in extreme poverty with no hope of escape. The exploitation I tracked took place in the fields, stone quarries, and brick kilns of rural India. This industry is widespread and victimizes millions. Why does this problem persist into the present? In interview after interview, I heard stories about how little local officials cared about enforcing the law. These reports of corruption and indifference line up with Bales' research. He finds that poverty and corruption are two of the key factors driving exploitation.[31]

This corruption is pervasive at the grassroots, but endemic throughout the official systems tasked with keeping track of brick kilns and their operation. As a result, no entity in India manages a comprehensive list of the number of kilns, to say nothing of the rate of on-site exploitation. Researchers and activists who want to target these spaces for research or interventions are left to their own devices, working at a painstaking pace across an unmanageably large swath of the subcontinent. A population-level assessment of the number of kilns would be a boon to both scholars and India's nascent anti-slavery movement.

Strong associations of advocacy groups have mapped out these locations terrestrially, but this is a slow and arduous process. At the University of Nottingham, our Slavery from Space project has set out to change that.[32] The project draws on the experience of Doreen Boyd, an earth-observation scholar, and Stuart Marsh, a geospatial engineer at Nottingham's Geospatial

Institute, as well as the expertise of Planet Labs, a private earth-imaging firm. At Nottingham, Kevin, Stuart, Doreen, and I are part of a unique interdisciplinary Rights Lab focused on path-breaking research on slavery and emancipation. This includes rights-focused earth observation that draws on a combination of satellites, crowd sourcing, and artificial intelligence to scan satellite data and create the first register of brick kilns in India. Future applications could include documenting the scope of the fishing industry in Ghana or charcoal harvesting in the Amazon. Exploitation is rampant in each place.

This process got its start as a crowd-sourcing initiative. The idea was to recruit volunteers to analyze photos from Google Maps and tag areas that looked like brick kilns. When this pilot was successful, the team expanded its efforts to machine learning. Of course, none of this will be of any use without a vibrant and robust network on the ground. Better views of old problems can only complement new solutions. This is the point made by observers like Jakub Sobik, a spokesperson at the UK-based advocacy group Anti-Slavery International, who noted that there are "more pressing challenges like ... withheld wages, lack of transparent accounting, [and] no enforcement of existing labor laws."[33] For those familiar with the struggle for indigenous, women's, and Dalit rights in India, these are perennial concerns for which there must be democratic solutions. Satellite imagery offers a key pressure point in international advocacy efforts, giving civil society groups leverage in shifting behaviors of elected officials and bureaucratic officials alike.[34]

The use of satellites to identify potential sites of exploitation is not emergent—a functioning and honest government could develop this assessment using far older technologies, like bureaucracy and the social survey.[35] Whether this use of satellites is disruptive cannot be answered so easily. If by disruptive we mean *the use of a technology to do something politically or socially unacceptable,* and by non-disruptive we mean *the use of a technology to do something that is acceptable to dominant political and social norms,* then we must determine what constitutes social and political acceptability in any particular case.

The critical question, then, is *why* the Indian state has failed to develop a system for auditing and inspecting sites where labor rights violations regularly occur. It is widely recognized that it is low caste community members who are the most vulnerable to severe exploitation in India. Why has casteism, like racism in the United States, proven so durable? There is a convincing argument that both represent widely held social and political norms. If this is the case, then the use of technology to illuminate the prevalence of exploitation in a political space where the rights of the marginalized are regularly violated *is* disruptive. The logic of this argument stands, whether in relation to caste in India, as satellites map kiln operations, or race in America, where drones could follow abusive law enforcement officials on their beat. At the Slavery from Space project, our use of satellites is non-emergent and disruptive. It could be done using lots of other tools, but it isn't.

Until now.

4 THE CAMERA'S POLITICS: HOW TECHNOLOGY TAKES ROOT AND TAKES FLIGHT

Several years ago, I was approached by community activists working in an out-of-the way village in rural Hungary.[1] The group served a Sinsi-Roma community living in a neighborhood affectionately called Miskolc, or "Numbers Street." The Roma are discriminated against across Europe and often live in poverty.[2] As we drove out to the site, Sandor Szöke explained the situation to me: these folks have been living here for almost 50 years, during which the community kept to itself and was generally left alone. The shift from Communism to a free-market economy had a certain effect. Nevertheless, subsistence living continued as it had in the past.

While Hungarian society has never been welcoming to Roma communities, a recent wave of pro-nationalist and anti-immigration sentiment threatened the community's existence. These dangers are multifaceted. Economic threats take the form of labor-market discrimination, cultural threats take the form of educational segregation and exclusion, and physical threats take the form of direct acts of violence.

We were on our way to Miskolc because, as if to add insult to injury, the local mayor had decided to expand the local soccer stadium by building a parking lot directly over the Roma neighborhood, effectively overwriting the group and its history.[3] Since there was no legal way to secure the land from its Roma inhabitants, a campaign of destruction began against its residents. Szöke recollected, "At night, thugs would come and strip out the windows, window sills, and door frames of unoccupied houses. Then the city inspectors would come and report that the place was abandoned, and thus mark it for demolition."[4]

These tactics were carried out in residential areas, triggering one resident to commit suicide out of despair. This suicide galvanized Szöke and a local advocacy group working with the Roma community. A glance at the site clarifies the soccer stadium's proximity to Miskolc. Yet the stadium is also right next to a large field. *Why not just build the parking lot on the empty field?* Szöke wondered. The solution seemed simple enough, though the activists faced a challenge: how could they raise awareness among Hungarians—and the international community—of a complicated issue in an out-of-the-way place? The campaign of violence occurred at night, by anonymous parties, and with no clear connection to the nationalist mayor.

Walking through Miskolc, I could see the broken windows, the lintels and sashes torn away and the interiors gutted. State-sponsored violence was being directed toward a historically marginalized and legally protected class of Hungarian citizens. Yet, to the untrained eye, Numbers Street looks like yet another broken-down post-Communist village, with linden trees expanding into the spaces where people had once been. From my perspective on the ground, the story could not have been harder to tell.

But I was in the village because Szöke had found a solution—a new way of seeing the problem and of telling Miskolc' story. In the dead of night, he had snuck onto government property with a pickup-truck load of small stones and a co-conspirator (who declined to be identified out of fear of retribution). Together they placed the rocks in a singular pattern. Szöke's logic was that if the mayor wanted a parking lot, he could simply use the adjacent field. The patterned rocks were intended to signal this possibility: together they piled the stones in the shape of a large parking sign, inscribed in the field, visible only by air.

Together, we worked on a video that combined drone footage with the activist's overview of the problem, and their simple solution. As the video pulled out from the parking sign and panned from the stadium to the Roma community, it was quickly clear to any viewer that the expansion was less about additional parking than it was intended to purge the city of a longstanding Roma neighborhood (figure 4.1).

Figure 4.1
Park here (author photo).

This disruptive land art follows in the footsteps of artists like American sculptors Robert Smithson and Michael Heizer. Most scholarly assessments of their large outdoor art (discussed in greater length in the next chapter) have focused on the extent to which their work was a rejection of the constraints of the New York City gallery scene and their concomitant embrace of the American West—or their idea of that West. They are also representative of the moment when artists made work with the knowledge that it could—would—be seen from space. Where a generation of scholarship on land art has focused on what is on the ground, the ability to actually appreciate this art is the result of photography from the air and the consolidation of those images into gallery exhibitions and books. The act of creating objects for visual consumption is a product of their visibility, and land art like Szöke's only makes sense when cameras can be taken into the air. We take aerial photography for granted, and thus far I have emphasized the importance of new aerial *platforms*, regardless of payload. In this chapter, I turn our attention to the camera itself, ask where it came from, and suggest some things about where it might be going.

FROM PHOTOGRAPHY TO THE CAMERA

The slide-lantern show was a groundbreaking tool in the struggle against the transatlantic slave trade.[5] It was a new way of seeing in both senses: visually with the naked eye, but also viscerally with a pricked conscience. The lantern show exposed viewers to the reality of slavery. Film from the liberation of Auschwitz had a similar effect—of forcing viewers to see two things at once: emaciated survivors and systems of annihilation. Viewing such images, advocacy writer Sam Gregory suggests, generates a co-presence for good that draws people together and spurs action.[6]

But how do we establish co-presence? Artists and movements deserve tremendous credit, but what of the gear itself? While much has been made of the role of photography in human rights advocacy[7] and social movements,[8] less has been said about cameras themselves. How, when, and where image-making can occur (i.e., where cameras can go) are critical questions as activists and insurgents square off against the state and other systems of authority.

In surveying the progression and development of geospatial affordances, it becomes clear that airborne cameras represent one of the most significant and potentially transformative payloads a satellite, kite, balloon, or drone can carry. While this volume has focused on geospatial affordances as *platforms*, in this chapter I focus on one payload in particular: the camera as a sociopolitical technology. Together with the novel, images of injustice have ushered in an era of humanitarianism that continues to this day. This is widely noted, but in that literature the advances are credited to the photograph, rather than the camera.

Photographs are celebrated. Cameras are ignored.

Inattention breeds under-theorization. As devices, cameras are often addressed obliquely, through their progeny, the photograph. When cameras are discussed, it seems they are only known by their fruit. The image is often the only thing about cameras that is of cultural and political importance—the means and mode of the photograph's production disappear into the background. Yet cameras are a technology with a history all their own. They do not simply exist; rather, their meaning is political, cultural, contested, and in flux.

The camera predates photography. There is something important about this fact: we built tools for relating to light (*camera obscura*) prior to the development of processes for making the relationship permanent (*collodion albumen*). Our contemporary history of the device as an actor in social space is often traced back to a horse-drawn apparatus that required a wagonload of chemicals. Its form was later condensed into the Speed Graphic, Rolleiflex, and Leica cameras that made street and war photography possible. Today, cameras have eschewed any singular form and are rather an array of sensors behind a lens. The singular consumer experience of the camera is as likely to have been replaced by something that is not at all a camera, but is in fact a small slab of screen that does myriad things. The lens, in fact, is the only component that remains unchanged.

This may not seem interesting to people accustomed to taking pictures first with Polaroid and 110 and later 35mm, and taking films first with 8mm video and later VHS cassettes. In these transitions (from 110 to 35m or from 8mm to VHS) the thing got smaller, but the underlying principles remained intact. The device gripped in the hands shrunk in size, had more features, and cost less money. With the shift to digital imaging devices, storage, manipulation, and reproduction became virtually free. The digital took our images from film and print to sensors and screens. Yet, in this retelling, I have shifted almost subconsciously to the image, the photograph, the thing produced by the device. We can take more pictures, we can store more pictures, and pictures are cheaper to have and to share.

Pictures, pictures, pictures. Such is the edifice of the image.

In the pages that follow, I want to worship instead at the altar of the device, to focus on how changes in the means of (image) production have led to changes in the social and political act of image-making. We used to press industrial tools to the eye. We now hold the world at arm's length and gaze at it until we are happy with what we are about to make of it. Such is the age of the selfie—we have come to look at ourselves in the same way.[9] Gone, perhaps, is Roland Barthes' notion of the punctum—the photograph's prick—replaced by the pout of the self-assured. If new technology has multiplied the amount of our world that can be imaged, it has also undermined the photographic moment.

The civil contract of photography relies on a sense that a photograph is being made.[10] This sense is no longer as obvious as it once was. Gone is the era when the act of bringing a camera to the eye was an obvious and political gesture that signaled a particular event—often personal, sometimes political, never nothing—was about to commence. Even the camera at rest in moments of political uncertainty represents a potentially hostile gesture to the powerful and powerless alike.

This line of thinking brings us back to the nature of the device in question. The merit of geospatial affordances rest in their ability to extend the visual, the public's line of sight, beyond earlier limitations. If we were searching for an analogy, perhaps we can settle on that of a camera-delivery system. While geospatial affordances can carry myriad payloads, it is camera-equipped drones that generate the most attention and concern. This concern is related to our privacy—who wants to be spied on by a disembodied and decontextualized eye, an unholy union of the male gaze and the Platonic view from nowhere! The concern is also connected to the importance of one's ability to see old things from new perspectives and to see new things altogether.[11]

What, then, is the nature of this turn of events? Answering this question requires a new line of thinking about the tools and tactics of photography (if not the photographs themselves).

New technology always attracts new attention—observers were in awe of the first daguerreotypes of war, and Robert Capa was widely heralded for his images of conflict. Once again, it is the image that has received the attention, rather than the underlying technology. It is too much to say that the camera has no sociopolitical history, but it is worth asking whether the camera having a history matters at all for anyone. Photography itself is a rather minor area of scholarship, if we take the number of citations of leading thinkers as an indicator. Cameras are the product of their technological milieu. This is, after all, why daguerreotypes lived between painting and chemistry, between art and science. That is also why the 35 mm camera shares so many design elements with other built and bored metal machines, including the Krupps canon and the rifle scope. And this is why new digital devices, cameras included, live within an ever-growing ecosystem populated by other chipped sensors.

Cameras are a product of their time—certainly when it comes to the means and mode of production, but also in terms of their relationship with social and political norms. It matters a great deal for politics and society whether a camera takes an hour to set up and the resulting image must be processed on site, as was the case with Matthew Brady's photographs in the middle of the nineteenth century. This is quite different from the ability to shoot and store roll after roll of film as one advances with a military expedition, as was the case for Robert Capa. It is also quite different, socially and politically, if it is not just trained photographers who are specialists in their field, but instead everyone with a smartphone in their pocket and an uplink to the cloud that can speak out regarding political, economic, and social issues in real time. This puzzle gets even more puzzling if the camera flies out of human hands altogether. How does the view from above afforded by new camera technologies change the way we think about human rights, the environment, our cities, and ourselves?

CULTURAL CRITICISM AND THE CAMERA

Cultural criticism of photography from the likes of Walter Benjamin, Roland Barthes, and Susan Sontag has tended to focus on the photograph and its cultural significance. For postmodern scholars, the technical act of photography was the pinnacle of modernist hubris. New work from essayist and critic Susie Linfield demonstrates that the tendency for prominent cultural critics to merge emotions and analysis simply doesn't apply to noted critics of the photograph like Charles Baudelaire and Susan Sontag.[12] For the skeptics of photography—Linfield places Baudelaire, Sontag, and Benjamin in this camp—a broad concern about the ends is an oblique condemnation and critique of the means. It is the lens' affectation of truth that rankles so. This is not to say that the camera has been overlooked completely. Roland Barthes confessed: "The noise of Time is not sad: I love bells, clocks, watches—and I recall that at first photographic implements were related to techniques of cabinetmaking and the machinery of precision: cameras, in short, were *clocks for seeing*, and perhaps in me someone very old still hears in the photographic mechanism the living sound of the wood."[13] Later, in summoning his book's name, he argues that "it is a mistake to associate Photography, by reason of its technical origins, with the notion of a dark passage [*camera obscura*]," suggesting instead the term *camera lucida*, to emphasize not the device but the moment it produced, the moment of seeing.[14] Here we have, in two brief moments, the book's namesake: first as nostalgia and last as portal. In neither case is the camera treated on its own terms, as a tool in use.

The eminent cultural theorist Susan Sontag, for her part, evokes the camera more frequently, but with a good deal less nostalgia. Within capitalist society, the camera has "twin capacities, to subjectivise reality and to objectify it" in order to "define reality in the two ways essential to the workings of an advanced industrial society: as a spectacle (for the masses) and as an object of surveillance (for their rulers)."[15] Perhaps more insidiously, handheld video technology made the *private narcissism of self-surveillance* possible, further subjugating private lives to powerful interests;[16] like cars and guns, they are *active fantasy machines* and *fantasy weapons*,[17] and photography is sublimated murder.[18] Sontag's *On Photography* is best remembered for being about photography, but it does credible double duty as an epistemological and ontological salvo against the camera as both metaphor and machine.

Sharon Sliwinski is a social theorist whose work on photography has greatly expanded our appreciation for the way image-making informs human rights advocacy. This undertaking, clearest in her excellent *Human Rights in Camera*, has been accomplished without reliance on or reference to the device itself. The picture, Sliwinski argues, carves out a distinct role for the viewer—images of rights violations create an opportunity for judgment. This judgment rests on a series of aesthetic encounters, and it is these encounters that produce the cosmopolitan conception of humanity that we today mistakenly trace to the Universal Declaration of Human Rights. It is the image's ability to elicit normative critique that matters. The argument is provocative, and rests on the analysis of pivotal images, from the 1755 earthquake in Lisbon to the 1994 genocide in Rwanda.[19]

Most writers on photography have little to say about the tools used in making the image.[20] Author, curator, and filmmaker Ariella Azoulay represents an important exception. Azoulay has advanced a theory of photography built on a "new ontological-political understanding" of the art, which "takes into account all the participants in photographic acts—camera, photographer, photographed subject, and spectator—approaching the photograph (and its meaning) as an unintentional effect of the encounter between all of these."[21] The globalization of camera technology has created many images, but also "a new form of encounter ... between people who take, watch, and show other people's photographs, with or without their consent, thus opening new possibilities of political action and forming new conditions for its visibility."[22] While Azoulay turns to social relations between people, it is with an eye on the materiality of the camera as the mediating technology.

This physicality is evident in Azoulay's assessment of Aïm Deüelle Lüski's avant-garde and experimental image-making.[23] In their exploration of *horizontal photography,* Azoulay documents Lüski's challenge to traditional "vertical" photography's approach to the world. Lüski is an Israeli artist and philosopher, rather than a photographer *per se*—although this would not be inaccurate either, it seems—and his signature intervention is the creation of imaging devices that free "the camera from its status as a means for something else and maintaining it as a participant in the event of photography."[24] The point is less whether the camera sees what humans see or mean to capture, but that the device captures what actually is (readers familiar with the work of Bruno Latour may find this intervention compelling). One of Lüski's devices captures three horizontal planes, fusing with the scene and undermining the camera's objectivity—all in an effort to answer the broader question: *what is there?*[25] Lüski's cameras are perhaps more philosophical and artistic interventions than optical devices—"the cameras do not satisfy the desire to see. Instead, they invite the observer to lose the external point of view at which the human and camera eyes are supposed to meet," Azoulay writes.[26] In this way the devices are explicitly *not* about the photographer's narrative. Neither are they about capturing the world, or a scene, as seen or framed by the photographer. The result is new sets of relations between the camera and the world, sets of relations that stand apart from the photograph and the viewer.[27]

For Azoulay, "participants" matter: camera, photographer, subject, and spectator. In a refreshing rejoinder to the critical status quo, the photograph and those who view it are out of frame altogether. All that matters is the camera and those in its vicinity. Azoulay's approach is so radical that work must be done to re-incorporate the image produced by this process and the observer of the photograph (distinct from the "spectator" of the original act of photography, suggested by Azoulay). This disaggregation—of the camera, photographer, subject, and spectator from the image and viewer—has important implications for how we think about the camera. In the pages that follow, I will suggest Azoulay and Lüski are on to something: the camera's origin predates, and its future extends beyond, human agency.

This argument takes the form of an interpretive genealogy.

THE CAMERA: A SOCIOTECHNICAL GENEALOGY

Image-making technologies have long been central to our under-standing of human dignity. Sharon Sliwinski argues that images are not just an illustration of ethical and political issues, but also constitutive of them. In this sense, she claims, "the conception of rights did not emerge from the articulation of an inalien-able human dignity, but from a particular visual encounter with atrocity."[28] If images are constitutive of what gets recognized as a human rights violation, then the camera device itself is an impor-tant technological artifact that shapes how issues are imagined and framed as objects of movement attention and action.

Shaky vertical video, for example, has become associated with the handheld cell phone camera and a sense of presence at the site of an event. This evidentiary vernacular is so widespread that a Norwegian filmmaker managed to fool journalists and human rights advocates with his "Syria hero boy" video that mimicked this style to tell a seemingly true story.[29] Drone video showing a bird's-eye perspective has become associated with the documentation of social movements and protests like the Umbrella Revolution in Hong Kong and anti-Internet tax pro-tests in Hungary as well as rights violations and environmental degradation worldwide.

Many of these examples demonstrate what Gillian Rose calls the *technological modality* of visual media—the technological apparatus that facilitates the making and displaying of images—and how they shape visual meaning making.[30] In this sense, Sandra Ristovska argues, "as technologies shape the material relay of knowledge, they are intimately connected to the ways in which the public learns about and remembers atrocities."[31] In other words, visual technologies are centrally implicated in the construction of rights knowledge. Invoking my experience with Sandor Szöke helps clarify this further—our footage captured the broken windows of the homes, the blighted context of the village, and the immediate proximity to the stadium. In so doing, we highlighted the violence directed against Sinsi-Roma homes and lives while underscoring the fact that this abuse is situated within a broader sociopolitical, economic, and spatial context: specifically, the humble community's proximity to the important cultural symbol of a football stadium.

How do innovations in camera technology, as a tool of documentation and witness, facilitate different ways of understanding human rights? Turning attention to the camera device itself, I argue, requires an eye for both the past and the present, perhaps with a sidelong gaze at the future. The story of the camera begins with a piece of architecture, specifically the notion of the camera as a room. The ability to capture light—to pin it down, to wed fleeting image with permanence—lagged by a millennium and a half. In what follows, I explore the evolution of the camera from its association with foundations and tripods to hands and sky.

Foundations—The earliest cameras on record trace back to the camera obscura—a dark space with an opening at one end, through which the light passed, and a flat surface at the other, upon which the inverted image landed. Kaja Silverman finds Mo Ti, the Chinese philosopher, musing about the "image-making properties" of such an approach as early as the fifth century BCE, followed by Aristotle a century later, and the Arab scholar Alhazen 1,500 years hence.[32] Many other thinkers spent the years between the eleventh and nineteenth century exploring the implications of this technology.[33] One important lesson that can be drawn from this origin story is the simple fact that, as Silverman observes, the technology stood on its own, figuratively and literally:[34]

> Since the viewer had to enter the classical camera obscura in order to see its images, he [sic] was also a receiver. This would have been hard to ignore, because the device had no focusing mechanism. The only way the viewer could render its often hard-to-see images more legible was to move around the sheet of paper on which they were received until he found the point at which they came into focus—i.e., to participate in the reception process.

What I want to emphasize here is not the metaphorical foundation the camera obscura laid for later photographic technology, but rather the materiality of an actual foundation.

In its earliest days, the camera was a *place*, rather than a thing. It was a fixed space where light streamed constantly, should an observer care to look. The light did not care either way; it came and went as it pleased, with nothing to hold it down. As a form of architecture, the camera obscura served artists rather than activists. I have found no record of any use or moment in which the camera obscura was directed toward matters of political consequence. Quite to the contrary, Sarah Kofman argues that theorists like Nietzsche saw the camera obscura less as an object for recording, but instead as a "metaphor for forgetting,"[35] forgetting being a kind of anti-politics in which information is acknowledged and then retired. The ability to permanently fix images as daguerreotype, and the invention later of glass slides that allowed for mass reproduction, ensured a permanency that allowed the image to pass easily into the political world, whether as a slide-lantern show or as a postcard passed hand to hand. The camera obscura did no such thing.

Indeed, the earliest camera artifact presents an opportunity for reflection rather than reporting.

Tripods—Technology shapes witness. Much has been said of Matthew Brady and Timothy O'Sullivan's images of the Civil War,[36] but the reality is that they are the product of a particular set of technological opportunities and constraints, which in turn shaped how the war was visualized. The shutter speed technology of the era required legs (a tripod), and at its earliest stages the chemical process of development required an entire wagon for its transportation. The camera was a device that could neither stand on its own nor be held by humans. These large cameras define an era of advocacy photography that stretched from the Crimean War of the 1850s and American Civil War of the 1860s on through the turn of the century, when lens and film technologies developed sufficiently to render the tripod unnecessary. The importance of these technical facts for image-making cannot be overstated. Images of war had to wait until the action ended, leaving ample room to situate and stage. Portraitures of generals were popular, perhaps in part because cooperative subjects could be made to stand still. The same can be said for the dead, who allowed the photographer sufficient time for composition. As Lawrence Douglas writes, "in an age of slow shutter speeds, the dead struck the most cooperative poses."[37]

The dominant technology—wet plate photography—required long shutter speeds and immediate film processing. Dark rooms were drawn on wagons and chemicals had to be mixed by hand.[38] Early versions of the process had to be completed—from coating to development—in as little as 10 minutes. Early dry plates eliminated the need for a tedious chemical process, but extended exposure times considerably.[39]

The images produced during the conflict made a huge impression on the public,[40] but the nature of the images themselves was shaped by technological constraints beyond the photographer's control and decidedly outside the viewer's field of vision. In sum, technology framed the nature and range of photographable moments: camera technology had a lasting impact on what was seen, recorded, and reported.

The era of the tripod marks photography's earliest relationship with humanitarian image-making. Roger Fenton's footage of the Crimean War in 1855—including his famous "Valley of the Shadow of Death"—was followed by what may be the first images of corpses, photographed by James Robertson and Felice Beato. While some of these images may have been staged—unsurprising considering slow shutter speeds and the influence of the theatre—their importance lay instead with the role they played in sensitizing the public's conscience to the reality of violent conflict and in creating what Azoulay has called a "civil imagination"—the civic practice of reclaiming civil power—around the suffering of war.[41] Here, photography costs the powerful something, contradicting official narratives of the glories of war, perhaps. This story about the power of the image is as credible of a human rights origin story as the novel.[42] It took investment in camera technology, however, for photography to fully matter for human rights. This is why George Roeder argues that due to "the technical limitations of early twentieth century photography, the most striking images to come out of World War I were written ones."[43] The camera simply wasn't up to the task of following the action. No wonder photographs were staged.

Hands—Faster shutter speeds facilitated the advent of hand-held photography and the emergence of a new ethics of authenticity. The slow speed and sizable entourage required by the tripod era meant that cameras occupied significant and visible time and space. There was no *capturing* an image in a traditional sense, only *framing* it. Faster shutter speeds liberated the camera spatially (from the tripod) and temporally (from the several-second wait it took to capture the image). This shift from the tripod to the hand had a significant impact on advocacy photography. Socially, faster shutter speeds meant the photographer did not need to secure the subject's cooperation to make an image, since one needn't ask them to stand motionless for an extended period of time. One only needed to *take* the picture. Imaging on dry plates—and later on film—also freed the photographer to make as many images as they had cartridges, without the need to process the film immediately.

At the turn of the last century, English missionary Alice Seeley Harris used the popular handheld Kodak Brownie to document Belgian atrocities in the Congo Free State. Sliwinski demonstrates how these photographs were used as an advocacy tool by missionary reformers but also as "a kind of forensic evidence of colonial brutality" in reports presented to British Parliament.[44] The images were meant to provide "incontrovertible proof" of atrocities and to inspire international humanitarian intervention.

These early rights photographs helped illuminate the fact that the Congolese people had been grossly abused, and in turn framed these abuses as criminal. The reformers conceived of rights in direct response to the suffering registered by the camera's lens, a form of compassionate responsiveness to that moment in which human dignity was thought to be lost.[45]

In other words, the Kodak camera contributed to a new organization of human dignity in international politics and the first articulation of crimes against humanity. It also inaugurated calls for moral responsibility premised upon the effective and evidentiary power of the visual. The use of these photographs by the Congo Reform Movement set the blueprints for subsequent humanitarian movements. The handheld camera became an inextricable part of the human rights advocacy toolkit.

The handheld era is also commonly associated with the rise of the Leica and similar high-quality, lightweight, cartridge-based devices. The gap between photographers like Jacob Riis and Lewis Hine and their successors Robert Capa and Sebastião Salgado was more aesthetic than technological—the style of their images and their image-making sensibilities may differ, but the underlying technological equipment used to make those images remained relatively stable. The emergence of the Leica and Rolleiflex in the late 1920s put light-weight and horizontal viewfinder-equipped roll film devices into the hands of generations of journalists and advocates, leading some to dub this the era of "Rolleiflex Witness."[46] Thirty-five-millimeter cameras fit easily into the hand and made on-the-fly composition easy. This format made possible Riis' humanitarian street photography, Capa's infamous work during the Spanish Civil War, Lange's frank portrayals of the dispossessed and interned, and the immediacy of Salgado's images of people going about their lives. We do well to remember these images, but I hope to continue to direct our gaze to the importance of the actual photo-making apparatus, of the equipment, in this process.

Rapidly interchangeable film canisters were eventually replaced by digital memory cards and film by sensors. Lenses, for their part, eventually evolved off of camera bodies and onto the backs of mobile phones, which are now the largest segment of devices in circulation. Yet from Riis' use of the flash (magnesium powder on a frying pan) on through the fast shutter speeds of Salgado's Leica and to a wave of handheld digital devices today, we see more similarity than difference. Mobile devices reverse the gaze such that it is no longer just the state that renders human rights claims legible. Mobile devices also enable activists to expand the horizon of what counts. What matters is that these devices are *mobile* and handheld, not whether they are digital or analog. From close range, the digital-analog divide appears significant. With a step back, however, we see that the era of the handheld device is but one important stage of several.

Generations of advocates were able to pursue their craft and bear witness in backrooms, alleyways, and battlefields the world over. What holds the handheld era together despite the digital/analog divide is the flexibility it provided to a broad range of photographers. Unobtrusive spontaneity was simply not possible in the era of the tripod. This unorthodox approach suggests that little has changed between CNN's 1989 filming of Tank Man in Tiananmen Square and recent viral videos of police abuse captured by citizen journalists. A lot has changed in the ways these images have changed hands, but little has changed about the relationship between the device and the human agent.

So what about handheld cameras and politics? Handheld devices have long been pivotal in depictions of rights violations. The modern human rights regime is firmly rooted in the global response to the Holocaust, itself extensively documented by both Nazi perpetrators and by the Allied troops who liberated the camps. Outrage led to new institutions and norms around universal human rights and set the tone for a new generation of camera use by human rights advocates.[47] I am arguing here that this is an era firmly rooted in an experience of the individual photographer, whose limber approach to image-making is facilitated by increasingly powerful cameras that allow images to be made on the photographer's terms. Tripods are helpful, but not necessary. Images may be distributed slowly (hand to hand via magic lantern shows), broadly (in newspapers or via postcards) or instantly (as with a live smartphone feed to a networked and watching world). Here, too, the radical transformation of cameras' *product* has attracted the most popular and scholarly attention. While the means of image distribution may have changed dramatically, the power of image-making has rested on the camera in the hands of the photographer. It is this era that geospatial affordances are now busy disrupting.

Sky—This selective and truncated survey of the history of the camera leads us back to the book's focal topic. If the earliest cameras were rooms or relied on tripods and human hands, the present technological moment is producing a new wave of image-making devices that have limbs and lives of their own. What form these entities take is subject to much debate, as scholarship on cyborgs and artificial intelligence makes abundantly clear.[48] Here, though, I am thinking specifically of satellites, balloons, kites, and drones and how they extend the threshold of visibility, contributing to different human rights imaginaries and advocacy opportunities.

Satellite imagery is an increasingly accessible tool for human rights advocacy. Lisa Parks has traced the use of satellites to document atrocities in Srebrenica as an example of efforts to "regulate the meanings of the war from orbit."[49] This meaning is managed by the state's selective acknowledgment of forensic evidence of atrocity, Parks argues, and in this way can be interpreted as an atrocity in its own right.

Here the footage indicts both the agent and subject of surveillance.

New work coming out of Google Earth raises the possibility of near-instantaneous event monitoring. Commercial satellites are able to capture ever-higher resolution images of the earth's surface in near-real-time, and small affordable satellites have put less-sophisticated imaging into the hands of non-state actors with a bit of technical know-how. Imagine the delight of journalists, police, scholars, and spies who may now receive an alert any time crowd density spikes in Tiananmen Square.

This is important information for scholars and advocates interested in public policy and civic engagement or concerned about the violent repression of civil society. Ubiquitous satellite imagery and selective imagery by balloon pose new opportunities. These opportunities are also available to states and corporations who want to enhance and expand control over information and resources. They also obtain for challengers, like insurgents, rebels, and protesters, who are interested in either resisting the powerful or in proactively securing new resources or information. While Google Earth provides ubiquitous satellite coverage for free, on-demand coverage is prohibitively expensive. Drones and balloons, therefore, are the technology that puts control of image-making into the hands of change agents. In a twist that will matter to only a handful of readers, the iconic camera company Hasselblad—founded in Sweden in 1841 and supplying to NASA the cameras that would make virtually all of the early iconic images of earth—was recently purchased from the private equity firm Ventizz. The buyer was DJI, the world's leading consumer drone company.

DRONES FOR SOCIAL MOVEMENTS
AND HUMAN RIGHTS

The camera's shift from tripod to hand to sky is significant for civil society actors. Small UAVs are able to provide a more constant stream of images than are available from satellites, and at much lower cost. Likewise, helicopters are able to do many of the things small drones can, but require financial capital for the craft and political capital for official access to airspace. Human rights advocates and social movements are very rarely in possession of these resources, making lightweight and easy-to-pilot quadcopters an accessible and affordable alternative. It goes without saying that putting a camera in the air makes new spaces visible. Walls and roofs and trees are no longer what they used to be. One need only think about the extensive efforts to harden American embassies after the attacks of September 11, 2001. A terrestrial glance at any embassy reveals a phalanx of physical obstacles, hardened guardhouses, and shatterproof glass. A view from the air suggests little thought was given to devices that could find their way into every nook and cranny, no matter the height or angle. The same can be said for skyscrapers, prison compounds, factory farms, prison camps, mass graves, plundered wealth, and any other secreted location relied on by the powerful.

An early implication is that the age of democratic surveillance is upon us. By *democratic*, I hope to signal the shift from the high-cost and top-secret tools used by the powerful to a more accessible set of resources used by everyone else. Camera-equipped drones do for the atoms of open air what hackers have done for the bits and bytes of the Internet.

Drones allows us to hack the world of atoms.

New devices allow us to gather new data and tell new stories, while also raising tricky questions about transparency and accountability. Who will hold a drone-equipped advocacy group accountable for footage of at-risk migrants? Will the rules be the same as those that currently apply to photojournalists or camera-equipped activists? Perhaps a new set of criteria will emerge. Will self-surveillance emerge as a pre-emptive tactic for actors challenging states and other powerful authorities? It seems reasonable to suggest that if police officers should be equipped with body cameras to look out at the world, they should also be tethered to drones that observe the officer *in situ*. If this seems like a radical proposal, we can then ask who is more likely to implement it—independent monitors like the American Civil Liberties Union, or police departments themselves, as a preventive measure? Critics may rightly observe that if police kill with impunity despite the ubiquity of mobile phones, then always-on drone surveillance might not matter. Indeed, the nature of the sociotechnical is such that technologies fit within larger sociopolitical realities. If stable repertoires of use are nowhere on the horizon, then laws and norms are even further off. If the powerful can commission satellite surveillance, as George Clooney has recently done over Darfur, then should The People be prevented from returning the favor, by for example using drones to surveil George Clooney's residence or, perhaps, taxpayer-funded military installations?

For the foreseeable future, these questions will be answered by evolving social norms, rather than laws, since it is not at all clear what new laws should cover, nor how new regulations will be enforced. The specter of autonomous drones takes this debate in a different direction, away from direct human agency.

Nevertheless, the privacy of the subject and the surveillance of individuals and institutions remain central. The technologies are new, but the tensions represented here are quite old. Camera-equipped drones, whether piloted by algorithms or humans, force us to confront anew the extent to which a particular vision technology is implicated in discussions about privacy and surveillance, control and resistance. In particular, the evolution of the camera has consistently unsettled whatever notions of space, time, and agency had evolved in an earlier era.

SPACE, TIME, AGENCY, AND THE CAMERA

Space—A palmed Leica meant anyone could be photographed from a horizontal perspective. Curtains and walls—traditional privacy techniques for shielding oneself from view—continued to do their good work. Aerial imaging devices, however, move the camera to new places. It looks wherever it likes. The camera's shift from the tripod to the hand was huge, as its meant passersby on the street were as vulnerable to being captured and photographed as were subjects in the studio (though with far less time to fix their hair and put on a public face).

On the street, perhaps, little has changed. At altitude, however, we see that previously private spaces—backyards, rooftop gardens, and penthouse suites—are no longer off limits. This fact has been widely noted by those concerned with privacy and surveillance. Such concerns are real, but a more critical analysis would suggest that drones merely expose the privileged to the everyday surveillance experienced by everyone else.

Invasion of privacy is a real and ongoing issue, but we should be sure to take in stride the new concerns of those with sufficient capital to have avoided earlier surveillance techniques through escape to private yards, roofs, and exclusive top floors. Surveillance is a longstanding issue for marginalized communities. Rather than being a new thing introduced by new technologies, such as automated facial recognition or unmanned autonomous vehicles, sociologist Simone Browne argues, *intersecting surveillances* are essential to undergirding and sustaining racism and antiblackness.[50] In this way, a host of tools—branding, runaway slave notices, and lantern laws—have been mobilized to surveil blackness.[51] Debates over surveillance are important, but we should ask broad questions about who is affected. To tip my own hand, I am more concerned about the state using drones to extend its ability to police communities of color than I am civilians using drones to spy on penthouse suites.

Geospatial affordances open new spaces for observation. While drones certainly impinge on the privacy of individuals in some instances, they also open new opportunities to hold the powerful to account. Clearly, the drone zone—zero to 400 feet above the ground—is a new frontier, perhaps a special area slightly beyond sovereign reach, an aerial extension of James Scott's Zomia.[52]

Time—The rapid increase in additional imaging devices—beyond the handheld lens—impacts image-making's relationship with time. In the shift from long shutter speeds on tripods to short shutter speeds in the hand, both have one clear factor in common: the presence of a human agent making key decisions. What should be photographed, and when, are critical. Image composition is a hallmark of both image-making and image-observing: Barthes' search for the punctum in an image is mirrored by the photographer's search for the right moment of light. The arty agentic instantness of that moment is being eroded by always-on video feeds, by intervalometers capturing images at regular points in time, and increasingly by remotely deployed devices that make images all on their own.

By de-momentizing (what else to call it!?) the moment of the photograph, subjects are denied the time to prepare. The shift away from the physical presence of a photographer under a hood or behind a viewfinder has social and political implications about the way that people feel they are living on Earth and on the street and going about their lives. In an earlier era, the camera was either taking a picture or it was not. The advent of handheld 35mm photography, especially on the street, meant that there was no particular moment in which the image was being made by the photographer. Furthermore, almost anywhere could be the place where one was photographed—on the street, in the alley, in the dark. If the handheld 35mm radically expanded the where and the when available to the human agent, independent imaging platforms have taken all three processes one step further, shifting not only our understanding of time and space but also of humans' role in the process. Both satellites and drones operate in their own time and often make images continuously, or at regular intervals, or at the moment they are prompted by a nonhuman agent like an algorithm. If time is relative for the image maker, it is also irrelevant for the image-making: infrared equipped sensors can photograph by night, eroding whatever benefits we may feel come from the cover of darkness.

Agency—"We have grown accustomed to thinking of the camera as an aggressive device: an instrument for shooting, capturing, and representing the world. Since most cameras require an operator, and it is usually a human hand that picks up the apparatus, points it in a particular direction, makes the necessary technical adjustments, and clicks the camera button, we often transfer this power to our look."[53] So argues art historian Kaja Silverman.

It may be time to transfer this power back to the device.

The third implication of this broad shift to camera-equipped geospatial affordances is that the act of photo-making is increasingly out of our hands. If images are being made according to logics and criteria that exist independent of human actors, then we must address new questions about agency. Drones are able to navigate toward and then hover around objects identified by an array of sensors, independent of direct human input. Likewise, satellites orbit the earth, making and sending images all on their own. In each case, the resulting images are warehoused in server farms, awaiting analysis by algorithms programmed to tease signal from the noise. The moment of the photograph is disappearing into a sea of always-on sensors. The role of the human in every stage of the process is also in decline. There is no single isolated time when that thing is happening, no one finger on the shutter release—no held breath for the moment and release in an act of agency.

To date, the actors in our histories of photography have been human, directing the camera's gaze, pointing and shooting, focusing and view-finding. Human agency is built into the design language of the device itself. The outside of the camera is made for the human user—knurled knobs for the grip of a hand and dioptric viewfinder adjustments for the human eye. By contrast, satellites, kites, and drones are integrated into human-built networks and engage with human-initiated and -mediated tasks and processes, but often do much of their work on their own.[54] They are sent out by humans into other places—streets and skies and space—to see what they will, then report back to humans, or not. The near future will see the emergence of a new class of cameras that are deployed in response to data events. The moment of the photograph is no longer linked to the index finger, the plunger, and the eye.[55]

This is true for aerial camera platforms and always-on surveillance systems, but new technology is changing *what* the devices in our own hands capture and *when* this capturing occurs.[56] Google is using the hand's natural shakiness—earlier a liability—as an asset, since it provides the nano-variation in perspective needed to develop richer digital data. From there, its systems use machine learning (a convolutional neural network deployed on TensorFlowLite) to analyze digital images on-the-fly. The result is critical determinations about coloration and contrast, but in low-light conditions it also results in the generation of details that are imputed algorithmically rather than captured optically. In other words, according to Google's Isaac Reynolds, the process increases "actual resolution so we can take pictures that resolve better than the underlying sensors might." The point here is stark: what the sensor captures and actual resolution are two different variables. Intelligent photography, Reynolds suggests, provides better pictures "not because it's helping you take a better shutter press, but it's helping you choose the better instant."[57] The device owner is now less a photographer than an editor, choosing images made by another. What the camera captures is mediated by artificial intelligence, and what you see isn't necessarily what you get.

When a phone does its capturing is also evolving. The term "shutter lag" refers to the momentary delay between when you indicate you want to take a picture (by pushing a button or touching a screen) and when the image is actually taken. Google has gotten around this issue with a "zero-shutter-lag" solution on its phones, which begin taking pictures as soon as the camera app is opened. A steady stream of images are buffered until the user indicates which moment should be captured, at which point the devices saves a dozen or so images, which it amalgamates into a composite image. When, exactly, is the photographic moment, what Roland Barthes called the punctum? It is gone, disappearing in the algorithmic flow.

The rise in computational photography suggests the handheld era is reaching its own inflection point. Cameras on drone platforms may be out of our hands literally, but the cameras that we have here with us may be out of our hands metaphorically. These lessons only become clear when we direct attention away from the photograph, the image, the product, and toward the sociopolitical implications of the tool itself. This chapter is a case study of what a device-centric approach to movement artifacts and political tools might look like, were we to attend primarily to the tool itself and its implications. It is also my own love-letter to the camera, or perhaps a kind of farewell.

5 RESIST!: RESISTING TECHNOLOGY AND THE TECHNOLOGY OF RESISTANCE

"Blue skies smilin' at me / Nothin' but blue skies do I see," wrote Irving Berlin, and for good reason, too. The sky is our lung-filling metaphor for the eternal and the infinite. The sky represents our earliest and deepest evolutionary and spiritual horizons: from the very moment when we first crawled from the primordial cave to feel the warmth of the sunlight on our bodies, we fell on our knees in awe. From sun worship to space launch, the sky has been each of our own, intimate connection to the infinite.

Our language about the sky reinforces this idea.[1] We engage in *blue sky thinking*, suggest that *the sky is the limit*, ask for a *room with a view*, note *there wasn't a cloud in the sky*, and tell the next generation to *reach for the skies*. The sky suggests freedom from constraints. The sky is prominent in popular mythologies about the American frontier: *Give me a home, where the buffaloes roam, and the skies are not cloudy all day* and *The sky at night is big and bright, deep in the heart of Texas.*[2] This is as true in Montana and Tibet as it is in New York and Shanghai. Around the world, we all expect to look up and see sky.

We rarely think about any of this, of course. But we will once this space is converted into an artery of commerce, another vein in the global flow of things, another node in the circuit of power, control, and domination. We will wax nostalgic for the old days, when it was just us down here, and everything forever above.

We know this because of the fierce resistance communities mount against the flight paths carved by airports. We know this because of debates over the placement of power lines and wind-mills.[3] There is a reason wealthy progressive towns have buried their utility lines and placed restrictions on how far above the ground advertising can protrude. It seems we are much more sensitive to visual pollution than anything else. View-blocking windmills off a coastline may be perceived to be more directly invasive than deforestation or the pollution of groundwater. We tend to get less upset over coal-producing plants we cannot see than clean-energy windmills that block our view of the ocean. In Denmark, no building can peak higher than the highest cathedral steeple. Europe is symbolic in its desire for capital to kneel before crown and culture.[4] How is this accomplished? By giving the sky its due place in the order of things.

Our relationship with the environment has been lately threatened by pollution, both visual and audible. We have, over time, learned how to control visible air pollution like smog (whether we actually do so is another matter altogether). We cordon airplanes off above certain altitudes and within particu-lar flight paths. Homeowners complain about the sound of land-ing planes in the same way they complain about interstate noise pollution, and in the same way they complain about the sight of old electric pylons or massive wind turbines on land and at sea. But such NIMBYists need only glance up—past the hanging eaves and fences and power lines and consumer clutter—to see their way through to the eternal and the infinite.[5]

It is here, in this air, that drones fly.

If I have thus far relied on examples of how geospatial affordances can be used for the greater good, this chapter is about something else altogether: how people resist drones and how drones are used for resistance. I'll be the first to admit that this chapter is a sort of fool's errand. Technology changes, and policy makers struggle to keep up. Once new regulations wind their way through the court, the technology in question will have morphed, yet again. When I started this book, the data firm Uber was disrupting the taxi industry. When I was writing this chapter, Uber had launched autonomous vehicles. By the time I got around to editing it, the program caused its first fatality. As I finished the book, Uber had announced its expansion into flying vehicles. This chapter may be outdated as soon as it is printed.[6]

Drones' mobility creates opportunities to access new spaces. New spaces are then used for both sight and for action. This can be seen in the preceding chapter, as activists contest the government's claim on land through an aerial image of ad hoc land art. Immobile art was translated into portable agitprop when we captured it by air—making something new out of the inscription on *terra firma*. Drones transform old spaces into new canvases. This is true for fields. It is also true for the sort of hard-to-reach surfaces that graffiti artists covet.

What public policies should be adopted amidst all this action? Whose property do drones travel through? How high does your backyard go up in the air? Can I shoot down a drone flying over my family's farm? If you shoot my drone as it flies over *your* farm, and it hits me on the head, what might a jury say? Can the Super Bowl deny airspace to drones? Can airports? Can I wear clothes that help me hide from drone surveillance? What if I am hiding from a police drone? Answering these questions is not easy, but the broader terrain they operate in is testimony to human creativity and agency. Whether the issue is individuals reclaiming public space from the prying eyes of the drone through airspace denial, or activists using drones to reclaim public space through public art and graffiti in hard-to-reach places, the human spirit is at work. New technology creates new realities that may inflame resistance, or inspire new action.[7]

DISRUPTIVE POLITICS

Skeptical readers might view this chapter's arguments more favorably by adopting what James Scott has coined an *anarchist squint.*[8] Scott's pop-anarchism is a friendlier sort than that developed by anarchists like Mikhail Bakunin or Pyotr Kropotkin, but it is equally focused on reclaiming space from larger systems of authority and control. If the preceding chapters have emphasized the ways this technology might balance state surveillance systems and hold the powerful to account, then I hope in this chapter to celebrate everyday pushback to drones of any sort and the use of drones to resist power of any sort. Here we run the gamut from legislation and clothes to anti-drone weapons and norms. Scott's anarchism, then, is less about abolishing the state than it is unlocking emancipatory human agency through the small acts of resistance and disruption that make social and political evolution possible. Scott, in a 2012 *New York Times* interview, confessed: "Unlike the anarchists, I don't believe the state will ever be abolished." It is instead a matter of "taming" the state through the kind of lawbreaking and disruption that have always been crucial to democratic political change.[9] In this way Scott echoes the call from philosopher and politician Roberto Unger that humans recognize our truest horizons as social, rather than political or economic.[10]

This is true from the air and on the ground.

Along the way, we may find new tools for nonviolent digital disruption that radically increase the state's cost of monitoring and repression, retaining the moral high ground ensured by nonviolent tactics. This is as relevant to people on the ground reacting to the state's use of drones for targeted killings as it is to the use of drones by disenfranchised communities monitoring states and corporate actors. Drones put increased power into the hands of the powerful. But they also level the playing field for those traditionally excluded from power—thereby opening new horizons for political struggle.

What form shall this struggle take? Taylor Owen echoes Thoreau in his convincing argument for civil disobedience: "Governments exist because people have assigned them representation and they should not take obedience to the state for granted."[11] The state's job is to "earn the loyalty of their citizens by pursuing justice," and, when it fails to do so, "civil disobedience is not only natural but should be encouraged." This approach echoes that advanced by John Rawls in considering civil disobedience as "a public, non-violent, conscientious yet political act contrary to law usually done with the aim of bringing about a change in the law or policies of the government."[12]

Technologies are part of the broader terrain of political, economic, and social struggle. This has always been true of digital technologies. Owen's intrepid research has turned up evidence that the earliest instances of digital disruption involved a computer virus that targeted nuclear systems (WANK—Worms against Nuclear Killers) and that the first effort to crash a server with simultaneous digital pings (DDoS—distributed denial of service) was an act of civil disobedience.[13] This raises complicated questions, he argues, about what it means to be violent or destructive online, and asks, "is a DDoS attack an act of speech or more akin to smashing a window?" How might this logic extend to drones, as they invade private spaces with cameras, drop payloads in prohibited places, and spray paint illegally?[14] Who should adjudicate between uses in a time of normative and regulatory uncertainty? This chapter leans into those puzzles.

PUBLIC OPINION

There is no doubt that new technologies face a skeptical public. Drones are no different. An early poll from Monmouth University in 2012 found that 67 percent of respondents supported the use of drones to apprehend criminals, but that nearly as many were at least "somewhat concerned" about their privacy if law enforcement were to use the devices. Studies conducted after a boom in drone sales suggest curiosity outstrips concern, but that there is significant skepticism about both military and civilian use. The American public is uneasy about the domestic use of drones, and the UAV industry knows it. The *Boston Globe* dryly noted that companies making drones for the military are "struggling for acceptance" and are occasionally met with protests.[15] While resistance to new military technology follows one public opinion vector, resistance to civilian and commercial uses traces another. Public resistance to drones appears to be rooted in several distinct concerns related to recency, safety, and privacy.

Recency—A certain amount of resistance to drones doubtless comes from their very newness, independent of their potential invasiveness. There are many legitimate concerns and complex challenges to drone use, but some sturm and drang seems less related to thoughtful assessments of their actual impact then to a fear of the new. Assessments of actual impact merit additional concerns about safety and privacy.

Safety—Legitimate concerns persist in relation to safety on at least two fronts. The first is related to the airworthiness of individual UAVs. Some early models from major manufacturers demonstrated a tendency to forget their location and fly away from the operator. Motor failure can also lead to a loss of control and a rapid descent to the ground, i.e., a crash. Over time, parachutes may be added to UAVs, while navigation and collision avoidance systems are being improved. The airworthiness of single devices is but one of several issues. Safety concerns also emerge from a general awareness that drones are flexible platforms for the transport of payloads of all sorts. They can easily be mounted with automatic weapons, poison gasses, and explosives, and these can be delivered with ease to areas previously inaccessible to earlier generations of fighters skilled in improvised explosive devices. While these concerns are related to platform and control systems as operated by humans, algorithm-controlled devices flying solo or in grids are susceptible to viruses and the unknown.

Only time will tell whether these fears are justified.

Those concerned by how drones *could* be used might be surprised to learn that very few of these dystopic uses appear to have been deployed in real-world conditions. An American teen who weaponized a quadcopter was tracked down by authorities.[16] The terrorist operation ISIL has strapped bombs to several devices constructed out of corrugated plastic and duct tape.[17] Likewise, drones have been used to deliver drugs to inmates in prison, and could easily deliver weapons in this same way. Despite these malevolent uses, the dataset introduced in the second chapter suggests criminal activities comprise a negligible percentage of total use.

The final safety concern is easily the most significant: thousands of hobbyists trying their devices out in backyards and parks across the country may be a hassle, but they are nothing in comparison to a world in which commercial drone use has grown to scale. The market for delivery drones in urban, suburban, and periurban areas is likely to grow, thereby increasing the demand for rock-solid collision avoidance systems that manage new air routes. It is unlikely that concerns over the technology's recency will dissolve into indifference. More likely is a scenario in which the benefits of drone networks are widely embraced, but intense policy debate centers on the perennial question of natural resources and the commons. Specifically: whose open airspace should drones occupy? Debates over where to place airports are contentious, as few existing communities welcome the noise and hassle that new airports bring. There is good reason to expect drone grids either map onto existing roadways, or operate at an altitude at which they will be seen as pointillist flows. The latter is unlikely, though, as this space is already occupied by larger craft.

Privacy and surveillance—A third public opinion challenge relates to privacy and broader concerns about surveillance. There is great reluctance to allow the state, corporations, or neighbors to invade one's privacy. Piloted devices are the vector by which sensors go mobile. The previous chapter focuses in on a single critical payload: the camera. For a hundred years, journalistic photography has had a symbiotic connection with the street. The most memorable photographs of violent conflict, social protest, and natural disaster have almost all been made by a person standing on the ground. The horizontal plane has been the most important space for both the perambulating human and the observant photojournalist—our eyes, after all, face straight ahead. Much the same can be said of most state surveillance and the increasingly common use of surveillance cameras in commercial centers.

Technology has always erased and redrawn the lines between private and public space. A lot of what happens in seemingly private spaces is not actually private, especially online and inside our homes and vehicles. This increasingly applies to our browsing habits and the data passively generated from devices: for instance, my phone's accelerometer telling my mobile carrier, or insurance provider, that I've not jogged in days. The emergence of drones and the proliferation of sensors challenge and expand our understanding of which spaces are even private. Ubiquitous CCTVs are one of the earliest examples of this proliferation, since they open sidewalks, parks, and other public spaces to sustained and archived monitoring by commercial interests and law enforcement. When the video feed from CCTVs went to magnetic tape, the puzzle essentially involved privacy violations from these two parties. When the feed went to digital archives, subject to hacking and scanning, the privacy issue metastasized as quickly as the footage could be shared. Digital archives of street surveillance footage, combined with facial recognition and behavioral software, are poised to push these privacy issues even further. Concerns over racist algorithms should be taken seriously.

If CCTVs invite constant monitoring of public space, UAVs push the boundary between public and private space, since camera-equipped UAVs move the line of sight from the street to the air. This simple shift effectively pushes public space and the public sphere from the sidewalk in front of a money machine to every stairwell, courtyard, rooftop, and garden in your neighborhood. These previously private spaces are newly subject to surveillance. Or have they now become public spaces, part of the public sphere? Should technologists, ethicists, and public policy professionals simply increase the number and type of locations that are now public, or has something more profound occurred?

While these observations may sound pedestrian, their implications are profound. Security and privacy policies address the prying eyes of the terrestrial observer, not the roving airborne eye of a small UAV guided by GPS waypoints while streaming video over a secure link to an operator sitting behind a laptop in a nearby cafe, library, or office complex. "Open air" and "free space" are no longer as "open" or "free" as they might have been. They are instead vulnerable to occupation by both atoms and bits. Cyberspace scholars suggest new technologies are pivotal in "radically restructuring the materiality and spatiality of space."[18] Whether this space is used for the public good, or as a means of state and commercial surveillance, is just the sort of dilemma regulators must face. Cyberskeptics fear the panopticon, believing a "society biased toward hierarchy and capitalism generates the entirely rational impetus for ... surveillance."[19] Others argue for a contrast between liberation and authoritarian technologies.[20] Where liberation technology is egalitarian, authoritarian technologies are "fundamentally hegemonic."[21] If Predator drone strikes in Pakistan and Yemen represent challenges to notions of sovereignty, non-military geospatial affordances represent fundamental challenges to the notion of public space and sphere—no small difference, anarchists like Murray Bookchin might hasten to add.

New technology creates politics and counter-politics. Period.

Drones are poised to disrupt the actually occurring material and physical space we inhabit every day. This applies to physical security as well as privacy. The walls and barricades protecting terrorist training camps, Occupy gatherings, and Davos meetings belong to a world of line-of-sight threats from paparazzi and pipe bombs. The United States has hardened many an embassy over the past decade. Reinforcements include motes, ramparts, walls, and bulletproof glass. Industry-standard protection against an explosive-laden truck is generally useless against a commercially available drone carrying a toxic chemical with an aerosol dispersant flying too close to an air intake. Innovation of this sort is a hallmark of asymmetrical warfare and operates in spaces architects and urban planners are increasingly forced to reckon with.[22]

LEGISLATIVE RESPONSES

As with most other aspects of this project, the scenery will change before the paint is dry on the canvas. A spike in drone flights starting in 2012 prompted policy makers around the globe to respond simultaneously to the opportunity that drones represented to their tech industries as well as the potential threat they posed to citizens. In the absence of early and decisive action from the Federal Aviation Administration in the United States, sub-state actors passed stop-gap regulations intended to curb use. The city of Poway—near my campus in San Diego and home to the assembly line that makes General Atomics' Reaper and Predator drones—responded to early flights of small UAVs near firefighters with a blanket ban. "We aren't opposed to this technology," Poway Mayor Steve Vaus told a member of my research team. "We're not trying to hamper it or hold it back. Just don't get in the way of emergency operations and it's all good." Vaus is not alone in his response.

We estimated that of the drone-related laws on the books in 2015, a third dealt with drone use by law enforcement (requiring police to obtain a warrant before using a drone), almost a quarter restricted or criminalized the act of unconsented surveillance, 15 percent restricted or banned using a drone to hunt animals, 12 percent expanded the power of state legislative taskforces to focus on drones and their use, and 10 percent criminalized the use of drones to harm others or to fly over critical infrastructure. It seems obvious that the wave of legislation passed or proposed in 2013–2015 directly followed an uptick in drone flights over the same time period. Yet many of the proposals in this period were rejected by state legislatures. This lack of progress in the passage of state-level regulations speaks to the ambiguity of what constitutes legal and acceptable use in the United States, as well as a general lack of coordination in the window between drones' increase in popularity and the absence, in that period, of a more comprehensive ruling by the Federal Aviation Authority. Within a year, some of these mandates had been thrown out in court, and clashes between early local laws and federal policies continue.[23]

My team's early analysis suggests most sub-state policies tended to focus on either civilian use generally, or specifically on the potential misuse of the technology by local officials. Several adopted legislation in early 2013 that focused exclusively on the government's domestic use of drones. These included restricting the use in court of information obtained from drones, as well as law enforcement's use of drones with any capability to harm, incapacitate, or otherwise negatively affect a human being. More pointedly, the city of Syracuse, New York, banned law enforcement and other official use of drones as early as 2013, citing the absence of a legal framework that adequately protects the privacy of the population: "Unlawful use or sharing of the data collected by drones would represent an unreasonable and unacceptable violation of individual privacy, freedom of association and assembly, equal protection and due process in the City of Syracuse and guaranteed by the First and Fourth amendments to the Constitution."[24]

Some cities, such as Northampton, Massachusetts, have called on the federal government to end drone surveillance and "extrajudicial killing by armed drone aircraft," and have drawn attention to drones marketed to domestic law enforcement that are "designed to carry weapons" as indicative of "a chilling message to the American people."[25] Our data suggests legislation of this type has waned dramatically, as discourse at the sub-state legislative level is now dominated by concerns about the proliferation of civilian drones. Some municipalities have restricted such civilian flights. A few cities instituted "peeping-tom" ordinances that banned civilian drone use, citing privacy concerns.

More recently, some have passed site- or day-specific bans relating to sporting events, the visit of a dignitary, or other large gatherings. For instance, in November 2015, in what has been pitched as a particularly comprehensive regulatory framework, the city of Chicago instituted civilian no-fly zones around airports, police departments, schools, churches, hospitals, and private property without the owner's permission.

Across these efforts, it is possible to identify three broad regulatory responses: free flying, total grounding, and regulated use. The term "free flying" describes jurisdictions that have either not passed pertinent laws, or whose laws place virtually no restrictions on civilian drone flight. "Total grounding" describes jurisdictions in which all civilian flight is prohibited. "Regulated use" describes jurisdictions where these first two types have been replaced by some combination of rules about the weight of the device, the height and location at which it can fly, the distance it can be from the operator, and the registration of the operators themselves. At the broadest level, regulations have set out to either protect citizens from the state or to protect powerful interests from citizens. By the time this book is published, these laws will likely have changed, but the spatial politics drones encourage are likely to remain socially and politically relevant for some time into the future.

It is entirely possible that drone policies at the local and national level are caught in something Langdon Winner has called *technological drift*.[26] Winner uses this term to describe the interregnum between a technology's invention and the moment in which a response is mobilized. It is in that span of time that changes take place too fast for responses that protect potential losers. Humans have always had a hard time seeing around corners. But it is not just foresight we lack. Once there is action, Hannah Arendt argues, consequences are boundless: "Action … always establishes relationships and therefore has an inherent tendency to force open all limitations and cut across all boundaries."[27] How the policy dust settles when it comes to UAVs is anyone's guess—the same can be said for the ramifications of and legal challenges to these policy decisions.

RESISTING DRONES

As I was finishing this chapter, repeated reports of drones flying near Gatwick Airport, one of England's busiest, grounded all flights over a number of days. More than 100 people reported a drone near the airport, and nearly all of those witnesses were deemed to be credible. A "Drone Dome" was on order from an Israeli weapons manufacturer, but had not yet been delivered. The Royal Air Force set up shop and deployed their own equipment in an effort to detect and deflect the rogue UAVs. When it was all said and done, 140,000 passengers on 1,000 flights had been impacted during the holiday season. Two drone hobbyists living near the airport were arrested, though they were released without charges. As this book goes to press, long after the incident, it remains unclear who was flying these devices. In fact, it isn't even clear whether any flight actually took place.

Newton's third law posits that for every action, there is an equal and opposite reaction. Drones generate responses that are both social and technological. This is not new. The airplane led to the anti-aircraft gun. Spam generated spam blockers. Peer-to-peer file sharing led to digital rights management by rentiers that monopolizes creatives' revenue streams.[28] Technologies are the site of struggle, especially in those areas where tools are deployed in support of contentious political activity. Legislative and technological resistance is to be expected.

Resistance also takes more invasive forms as efforts to thwart drones are multiplying. Efforts to curb drone flights generally fit into five broad categories: sensing drones, denying airspace to drones, taking control of drones, destroying drones, and hiding from drones. In a 2018 report on counter-drone systems, the Center for the Study of the Drone, a US-based think tank, found that more than 230 different counter-drone products are in some stage of development by 155 manufacturers in 33 countries.[29]

Sensing drones—An important initial issue in denying airspace to drones is knowing whether a drone is present in the first place. The tech startup Drone Shield offers a number of drone-detection platforms that sense UAVs' unique sound profile and alert the shield's owner. The firm claims that their "acoustic detection technology" is able to identify devices that cannot be picked up by radar, or that are flying by GPS rather than direct radio control.[30] The Gatwick example suggests stable industrial standard solutions are a long way off.

Taking control of drones digitally—Device detection raises new questions about how to respond. Passive systems are able to actively and consistently deny airspace to UAVs. One such solution is SkySafe, a San Diego–based startup that offers a weatherproof box the size of a deck of cards. SkySafe claims their unit is able to sense a drone, then take control of the device and force its landing. The group's website includes a demonstration of a simple app-controlled unit that instantly disables a popular drone from the manufacturer DJI. Presumably the non-beta version of the technology allows the offending drone to be landed safely rather than unceremoniously and unsafely dropped from the sky.

Where SkySafe targets drones' unique transmission signatures, lightweight and portable global positioning system (GPS) transmitters can disorient the legitimate GPS signals that originate from the satellites that most UAV pilots rely on to assist them in flight. The successful disruption of a drone's navigation system can transfer control to another person, or can simply untether the device from its geocoordinates, thereby causing the drone to fly away or crash. While GPS spoofing causes the disruption of a drone's operation, affordable software from a Ukrainian company alternatively allows unencrypted communication from military drones to be harvested with a common satellite dish and the program SkyGrabber. This approach does not disrupt the drone's operation, but does compromise the security of the data transmitted by the device.[31] Some technologists have suggested that this approach may be modified to essentially disconnect the drone from its operator—an important discovery, if true, since the process describes intercepting communications with large, fixed-wing UAVs like the Predator and Reaper.

Smaller platforms can be similarly hijacked by other drones that wardial the target drone for a vulnerability. Wardialing is the process of repeatedly pinging a device for an open port or vulnerable point of access and then connecting remotely to the host device. The software package then hijacks targeted systems and "redeploys them as autonomous infectious agents against other nearby drones." This software was first announced in 2012, and its author, James "substack" Halliday, later won a 2013 competition with his "virus-copter" project. As I write this some years later, Halliday's hack is still available as a free download on the software sharing platform GitHub.[32]

The research and development team at defense contractor Battelle offers a less-involved alternative: their DroneDefender is an AR-15 rifle modified to hold a radio antenna instead of a barrel. The non-kinetic device is billed as a "directed-energy unmanned aircraft system countermeasure" that jams a drone's radio and GPS signals at ranges up to 1,300 feet. It is unlikely this claim can be verified by the public, since the device

> is not, and may not be, offered for sale or lease, or sold or leased in the United States, other than to the United States government and its agencies, until authorization is obtained. Under current law, the DroneDefender may be used in the United States only by authorized employees of the Federal government and its agencies, and use by others may be illegal. Due to Federal regulations, this video is a simulation of the Battelle DroneDefender™ system. *It has, though, been successfully tested in Federal government-conducted field trials.*[33] [italics added]

While this technology may be proprietary, restricted, and of interest to state security forces, efforts to bypass it will be open-source. Encrypted drones that are hardened against such vulnerabilities are already in the works. Advances in machine learning and vision will eventually remove humans from the loop, instantly rendering many of these solutions irrelevant.

Taking control of drones physically—Other countermeasures focus on physical contact rather than digital disruption. With the release of a video showing trained eagles snatching drones out of the air, the Dutch National Police reached geek stardom faster than any police force in memory. Michigan Technical University's (MTU) Human-Interactive Robotics Lab has developed a "DroneCatcher," which they bill as a robotic falcon. As MTU faculty Mo Rastgaar told the *Washington Post*: "You can't shoot a drone that has explosives. And also, force landing, that is also not a good idea. So, probably a drone catching another drone" is best. The UAV is mounted with a cannon that shoots a net at other drones operating within 40 feet, ensnaring them for delivery to their police handlers, since Rastgaar insists he will only sell the technology to law enforcement agencies. This same approach has been adopted by OpenWorks, who claims their SkyWall 100 can effectively net drones at distances of more than 300 feet. The French company Malou Tech has proposed a drone-based system that mimics the process of dragging a net through water: a large drone flies a net through the air, ensnaring smaller drones.[34] If such technology increases in popularity, drone operators are likely to develop more sophisticated flight-based responses, likely including random and erratic flight plans combined with evasive artificial intelligence–based responses. In late 2016, a system piloted by a modest onboard Raspberry Pi processor beat a human pilot in a dogfight.[35]

Destroying drones—The crudest way to bring a drone down is to simply shoot it, a so-called kinetic interdiction. *Popular Mechanics* has conducted a bit of testing to determine the best firearm for downing a drone. After some trial and error, they determined a 10-gauge shotgun with number 10 or 12 birdshot is best for the job. Like all of the approaches surveyed here, the legality of shooting down drones is vague. A Kentucky man was charged with criminal mischief and wanton endangerment after shooting a UAV down. While the case was thrown out, it is but one of a growing number of challenges.[36] In New Jersey, a man who had shot down a drone was indicted on two charges of criminal mischief and the possession of a firearm for an unlawful purpose.[37] Terrestrial shotguns are only one of several permutations. Students at the Moscow Aviation Institute have built and deployed a shotgun-equipped drone capable of firing 10 rounds at another drone (or anything, really).[38]

Future developments in kinetic interdiction will likely target both airborne approaches (flamethrowers, missiles, and suicide drones) as well as land-based approaches (spears, bolos, and boomerangs). While these uses are in beta, the final nature of airspace denial is likely to lay further along one or more of the paths sketched above: taking control of drones digitally or physically, or destroying them altogether. For the time being, and in the future to come, another question bears asking: how to best hide from a drone?

Hiding from drones—The process of giving drones the slip is also in beta. Even the most basic quadcopter can carry sensors that capture visible light, near-infrared (IR) and forward-looking IR. Visible light sensors capture the same data as the naked eye on a normal day, or what my smartphone camera picks up under normal lighting conditions. Near-IR is the view seen through commercially available night-vision goggles. These come in two flavors: passive IR, which amplifies small amounts of light in order to brighten an image; and active IR, which, in contrast, can capture images in total darkness once a scene has been lit by an IR-emitting device.

Hiding from drones involves a number of straightforward hacks, depending on which sensors the platform is carrying. One can wait for bad weather to ground drones, since smaller devices have a hard time in high winds, dense fogs, and heavy rain. One could also avoid using wireless communication like a mobile phone or GPS, as their digital signatures may reveal one's position. This is especially true during armed conflict. One can break up and distort the image a drone sees on the ground by strewing broken glass or mirrors on the ground. Finally, mannequins can be used to confuse sensors. Hiding one's self from drones involves obscuring at least one of four factors: the body, the face, one's gait, or one's heat signature.

Body—The Amsterdam-based designer Ruben Pater received widespread attention for his Drone Survival Guide. Seen in figure 5.1, the guide is a simple two-sided sheet. On one side is a guide to common military UAVs. On the other is a series of tips for hiding. Trees are important as they represent some of the "best cover against the planes." Space blankets keep one warm while also providing a shield against heat-seeking infrared cameras.

Survivalist and "prepper" websites pick up where Pater leaves off, with ideas for day camouflage (trees, shadows, forests, netting) and night camouflage (hide in buildings and under trees and avoid using lights, which drones can spot from great distances). At the most sophisticated end of the spectrum is multispectral camouflage netting, which provides protection against forward-looking IR imaging devices. At the least sophisticated end is the simple advice to hide someplace warm. When ambient temperatures hover between 95°F and 105°F, the body's heat blends into the surroundings, making it difficult for infrared technology to clearly discern the human form.

Face—Clothing represents the most basic means of hiding one's face, though an artist has recently released an antidrone hoodie that's intended to shield against heat signatures and facial recognition. A number of efforts are focused on the

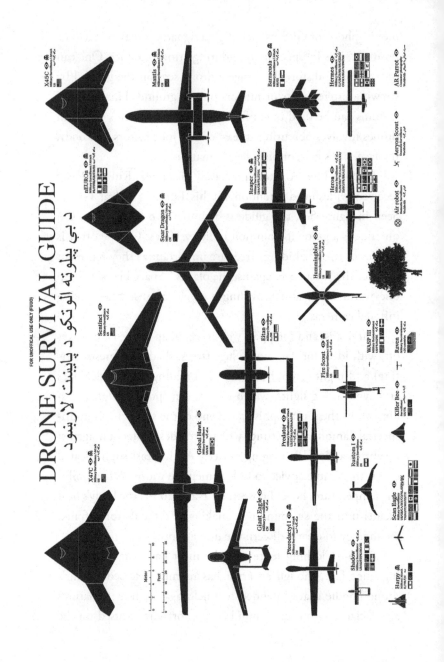

Figure 5.1

Ruben Pater, Drone Survival Guide (www.dronesurvivalguide.org; by artist permission).

development of glasses that will foil facial recognition systems.[39] More basic technology, like hats, umbrellas, and hijabs represent the kind of simple hacks that are likely to confound new technologies, at least in their early stages of development.

Gait—Gait-recognition technology relies on the fact that individuals have completely unique styles of walking. The lower body in motion provides a handful of discreet anchoring points of motion, which can be scanned for the unique combination produced by a particular individual.[40] Foiling gait-recognition systems may be possible by affecting a limp, adding body padding, or shrouding key pivot points under larger clothing, like a loose robe, or within more constrictive clothing, like too-tight jeans or a minor leg brace.

Identity and tracking technologies are likely to grow in new and unexpected ways. The dystopian British series *Black Mirror* sketches one such scenario in the episode *Hated in the Nation*, in which bee-replicating pollinator drones do double duty as the deep state's eyes and ears, enabling them to seek out people based on facial recognition software that is handily linked to the government's national identification system.[41] Humans are likely to develop technology to push back on such developments, as evidenced by the experimental techniques highlighted in this chapter. In another scenario imagined by *Black Mirror*'s writers, individuals are able to deny others the ability to see them. This on-demand capability centers on the right to be seen by, or blocked to, a particular person.[42] This capacity to toggle one's visibility is facilitated by corneal and auditory implants. Viewers watch a frustrated woman signaling the end of their relationship to her boyfriend by muting him visually and audibly. With a click of the button, he becomes a perpetually muffled blur, and she need not look him in the eye again. Future cloaking technology is likely to advance in this direction, perhaps alongside digitally enhanced contact lenses. More broadly, as drone technology changes, this sector appears to be in its infancy. Many of the systems described here are based on the radio-frequency signatures that connect drones and pilots. But as machine vision systems evolve, on-board collision avoidance and mesh flight paths may remove this link in the system, sending anti-drone innovators back to the drawing board.

For the present, however, a recent wave of legislation has set out to slow the spread of drones, new and old technologies and tactics allow some to hide from drones, and these together represent the leading edge of resistance to drone technology. Drones can also be used as tools of resistance and in support of counter-hegemonic actions intended to target the powerful. It is to these uses that we now turn.

RESISTANCE: ESTIMATING CROWD SIZE

Drones, satellites, balloons, and kites provide open and auditable data about events on the ground.[43] Though this functionality is not emergent, it can be disruptive, as it provides a transparent account regarding the size of an event. As seen earlier, social movements telegraph public opinion to leaders. In so doing, they bypass traditional political mechanisms, like voting or public opinion polls. Protest events provide new data directly to the media, political aspirants and incumbents, as well as the general public. Protests may put new items on the agenda or highlight the fact that seemingly settled issues are still of socio-political significance. While public gatherings of people are not the only way to garner attention, they have enduring efficacy in this regard.

Social media updates and opinion pieces may reach a particular subset of the public, but large events gain the attention of key opinion leaders in the economic, political, and entertainment establishment. It is not just event size that matters. The perceived worthiness of the cause—whether it is a call to protect the vulnerable or a claim to greater resources by the wealthy—matters. So, too, does the amount of unity demonstrated by those gathered at an event. Solidarity and camaraderie resonate in the public imagination. In the event a protest or mass gathering faces trials and tribulations of any sort—from incremental weather to abusive security forces—stalwart commitment to the cause matters as well.

This was on full display in Donald Trump's claim that his inauguration turnout was much larger than unofficial estimates and the naked eye suggested. The small turnout was amplified by the fact that the candidate had been decisively beaten in the popular vote. In this case, a widely shared comparison was made between the size of the crowd at Obama's 2013 inauguration and Trump's inauguration four years later. The contrast was clear and the implications stark, especially since Trump was clearly ignoring a reality visible to others. The image that made this possible was taken at an altitude and angle that only airborne cameras can capture in most public events. This is especially true in urban areas, where streets and sidewalks compress participants into auditable units.

Aerial imagery taken by balloon-, kite-, or drone-borne cameras are perfect for analysis by a number of industry-standard approaches. Of course, this data cuts both ways. I have had movement supporters suggest to me that a focus on crowd size directs attention away from the normative arguments made by the movement. The focus, they argue, should be on democratic claims rather than on a crass assessment of how many people showed up for an event. This critique may work in the ivory tower (which is indeed where it was made), but the political reality is clear: large events resonate with political incumbents. Determining whether an event was "large" should not be left to the police or protestors alone. Open-source crowd estimation will hold everyone to account.

RESISTANCE: MONITORING POLICE BEHAVIOR

While many human rights advocates have focused their attention on the weaponized drones used by governments and flown outside of the legal status quo and established international law, groups like the ACLU are instead primarily concerned with the police deployment of drones in American cities. This is an important issue. There is very little reason to believe that police forces in the United States should be trusted with the weapons they have, let alone new and more powerful devices.

Police departments can certainly find ways to put small UAVs to use, as evidenced by their exponential spread in the time I worked on this chapter. Drones can be used to monitor prisoners on probation, to identify crimes in progress, to identify and track criminals, to conduct high-speed chases, and, perhaps most distressingly, to simply loiter over areas where crimes traditionally take place and to wait for something to happen. This is the aerial extension of the big-data crime problem—that is, simply put: training data that uses past arrests to predict future crime simply reflect racist social norms and policing patterns.[44]

Individually, these technologies appear to be a step-wise extension of existing technologies: drones' hovering capacity is similar in type to that provided by CCTV; the deployment of UAVs to chase a suspect bears a strong resemblance to a helicopter dispatched to track a suspect; and the process of getting "eyes on" an individual recently released on parole is an everyday occurrence. Taken together, however, these uses can generate novel responses that lack accountability and accelerate existing abuses. It is likely that police departments across the United States, and around the world, will continue to work hard to secure this sort of command, control, and surveillance capacity, and in so doing will secure access to technology before underlying issues are resolved in favor of social justice.

Fortunately, police do not always get what they want. The American public, deeply distrustful of both the state and its agents, has made consistent efforts to push back on such uses. Liberals, conservatives, and the public more generally have reacted with some hostility to state surveillance. When the Seattle Police Department received a drone as part of the grant program, it was forced to get rid of it, shipping it instead to the Los Angeles Police Department, where it sat in storage, grounded by the same public pressure it faced in Seattle. Yet this grounding has not lasted long. Data on law enforcement's adoption of drones suggests sheriffs' offices and police departments are adopting small UAVs at a rate and pace that matches the exponential growth found in civil society. In a 2018 report, the Center for the Study of the Drone at Bard College estimated that 910 public service agencies were using drones, an 82 percent increase over the previous year.[45] The most popular platforms appear to be off-the-shelf technology from DJI.[46]

Civil society groups have responded in kind. The American Civil Liberties Union has taken the lead to address the privacy concerns of middle-class citizens as well as the Black Lives Matter movement's spotlight on systemic efforts by police systems to isolate, marginalize, criminalize, and tax communities of color. In other words, it is good news that so many are worried about the police having drones, and I hope their good efforts keep this technology at bay for the foreseeable future. This is not to say that there is no role for drones in police-civilian relations. Quite the contrary, there may be some merit in better exposing America's officers of the peace to more democratic systems of transparency and accountability. Two steps may facilitate this process.

Systems like Five-O and Excuse Me Officer allow residence to rank law enforcement officers in a "Yelp for cops," if you will—the idea being that communities can rate on a transparent and online platform their interactions with individuals from their local police force. Of course, the system may lead to some distrust and hurt feelings, but these will be growing pains on a path toward local and enforceable accountability for law enforcement officials. It is clear that democracies like the United States and France have systematically ignored their responsibilities in this regard, thus creating an opportunity for communities to begin a process of technologically enabled accountability at the local level.

The second step of this process involves enforcement, and suggests a possible role for UAV technology. Individual law-enforcement officials found to be in violation of community and legal norms could be placed on a type of probation that involves constant surveillance by individually tasked UAVs of all on-duty behavior. It would be the device's sole task to monitor and report on the behavior of the offending officer for a particular probationary period. Subsequent footage would be archived for periodic review by a panel comprised in equal parts of citizens and law enforcement officials until that point when the probation is lifted. Drones challenge the state's use of force; they also challenge dominant visual discourses relating to public space. This speculative use is both emergent and disruptive. A pause may be necessary to ask why the use of technology to hold the powerful to account is revolutionary in one of the world's oldest democracies. Here we are, nevertheless.

RESISTANCE: DRONE GRAFFITI
AND THE DECOLONIZATION OF SPACE

Graffiti is a critical artistic tool in the counter-hegemonic kit of the everyday anarchist imagined by James Scott. Unconventional public art challenges seemingly settled arrangements about the use of and control over space and, I would argue, is key to the expansion of the public sphere. The lines between public and private spaces are things of politics, and are therefore dynamic and open to critique. This is true of the zone between corporate and civic spaces. There is nothing new to this observation.[47]

Experimentation with drone graffiti, however, suggests an opportunity to take this critique to new levels. In many urban contexts, easy-to-see but hard-to-reach spaces have been occupied by those with sufficient capital to control their use. A prime example lies in the world of advertising firms and billboards. These spaces are harder to control if artistic media—spray paint, for example—is newly mobile. The artist KATSU made waves when he augmented a large Calvin Klein billboard in Manhattan by crisscrossing the model with red paint. The artist told

Wired magazine that the effort "turned out surprisingly well," despite the lack of precise control over the spray-paint-wielding DJI Phantom (figures 5.2–5.4).

The central objective in graffiti art is to get seen. As Cameron MacLeod, an early adopter of graffiti drone technology, explained to me: "The aesthetic of graffiti production is second to the thin red line that runs through all these subgenres: it's access and distribution. That's the bible."[48] It's not surprising then, that graffiti artists have long sought to find new places for their graffiti. Danish writer Jesper Vestergaard argues that drones have given graffiti artists new ideas about the things one can do with technology. While the broader graffiti community seems indifferent, a handful of artists are experimenting with drone-based tagging. These efforts appear to be distributed across two

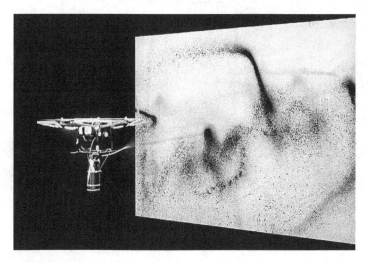

Figure 5.2
Graffiti drone, KATSU (by artist permission).

Figure 5.3
Drone graffiti (detail), KATSU (by artist permission).

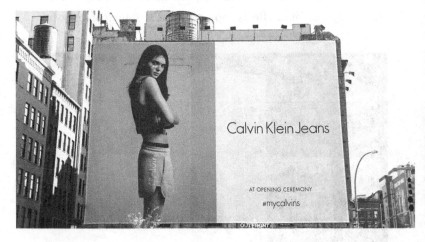

Figure 5.4
Drone graffiti, KATSU (by artist permission).

broad genres, the first related to artists using drones as a new form of expression, perhaps for artistic reasons, political communication, or both. The second is a cluster of hackers and DIY makers experimenting with the technology, including a dot-drawing project centered around McGill University in Montreal as well as the Flying Pantograph project at MIT. In explaining this broader context, Vestergaard suggests that the former group is more political, while the latter is more technical.[49]

For a smaller group of artists, the challenges posed by DOIP—drones over IP—is most compelling: "Being in two places at the same time is related to the oldest philosophical issues," Vestergaard told me.[50] More broadly, MacLeod points out, drones could "change the way graffiti is made, because it will give access to places there wasn't access to before." Indeed, the technical components of a particular graffiti tag are important representational practices. A team of highway taggers in Montreal put together an exhibition in which they stacked dummies atop one another to illustrate how they were able to tag particular areas of a busy underpass. Their installation emphasized the materiality of their artistic intervention.[51]

These efforts point us back to disruptive and contentious politics. "From a larger perspective," MacLeod explains, "it's about public control over public environments, and how that comes into conflict with [private] control." That political debate is more important, he suggests, than any particular attempt to define the graffiti drone as "good or bad." This observation is in keeping with graffiti's own legacy, as the art form emerged alongside emancipatory efforts within hip-hop culture in the 1960s and 1970s. Graffiti complemented broader efforts by disenfranchised youth to counter those media conglomerates "dictating the visual discourse." Graffiti drones extend this logic further, MacLeod argues, to a point in which "the underclass would have complete control over their aesthetic environment." For some, automated and remotely controlled tools are important technologies for realizing this goal.

As in other areas of use highlighted here, a graffiti drone community has not coalesced. Original trial-and-error efforts have morphed into loosely connected collective efforts to solve the complex technical problems involved in controlling a flying spray paint can. In fact, collaboration efforts have fallen afoul of competition, and no individual group possesses the resources necessary to solve such complex problems. What's needed, MacLeod tells me, is "somebody who really has a huge amount of funding behind them to do this… but there's not a lot of incentive from industry to create this either!" The regular use of drones to decolonize space, to expand the civil sphere, and to practice prosocial civil disobedience may never take off, but scholars should keep an eye on this emergent and disruptive use of new technology.

RESISTANCE: LAND ART

Nobody knows why the Nazca Lines in southern Peru were created. Several theories hold that the images of animals—including fish, monkeys, jaguars and humans—were religious symbols, possibly a signal to the gods that look down on the earth. One creative thinker has suggested they were instead the result of a civilization that had mastered flight.[52] Whether the observer was a god looking down from some magical altitude, or a human overflying the desert in a mysterious hot air balloon, there is little doubt the art was meant to be seen from great heights.

Built in a period falling sometime between 500 BCE and 500 CE, these are easily the longest-lasting earth-based objects visible from space. Stephen Graham has suggested that when the earth can be viewed from above, it "becomes a canvas for the work of artists and activists … to be consumed, via the Google Earth system, on the laptops and smartphones of a global audience."[53] Lisa Parks argues something very similar, suggesting in *Cultures in Orbit* that her book "is an attempt to wrestle the satellite out of the orbit of its 'real agencies' so that it can be opened to a wider range of social, cultural, artistic, and activist practices."[54] While we should rightly worry about the totalitarian state, Parks doesn't stop there, observing that "since the 1970s … media artists have been challenging the military, scientific, and corporate authority over these space-bound machines."[55]

Land art is a particular kind of challenge, as it turns land into canvas and pressures the viewer to stand at a great distance.[56] For Graham, this canvas belongs to artists like Raúl Zurita, the Chilean poet citizen who survived torture under the dictatorial regime of Augusto Pinochet. His poem—*ni pena ni miedo*—is a 3-kilometer-long installation in the Atacama desert, defiantly telling the world, or the universe, rather: *no shame nor fear* (figure 5.5).

The land art, or earthwork, movement emerged in the 1960s and 1970s, hard on the heels of Sputnik's 1957 launch and in the midst of a space race pitting the United States and the Soviet Union against one another. This movement is known for its rejection of New York City's gallery-centric art scene, but it is also an embrace of an art form that takes seriously a view of the earth—an approach to the earth—that can only be afforded by satellites and other geospatial affordances.[57]

Figure 5.5
"Ni pena ni miedo" (no shame nor fear), by Raúl Zurita.

The pioneering earthwork sculptor Robert Smithson died at 35 while overflying his seminal work, "Spiral Jetty" (figure 5.6). While much is made of the installation as a counter-site to the gallery, and importance is ascribed to actually *going to* and experiencing the piece, land artists are also critically aware of the land as canvas.[58] Smithson tried to square this circle by siting an installation near Dallas-Fort Worth Regional Airport, so that it would be seen by passengers in planes as they landed and took off. To the best of my knowledge, nobody ever asked whether this converted the pressurized cabin into a gallery from which the patrons (rather than the art) could not escape.[59]

The sculptor Walter De Maria—perhaps most famous for his installation "The Lightning Field"—proposed, but appears to have never actually executed, a "Three Continent Piece" that would have been "generated by satellite: three superimposed

Figure 5.6
Robert Smithson, "Spiral Jetty."
Copyright 2019, Holt/Smithson Foundation and Dia Art Foundation. Licensed by VAGA at Artists Rights Society, New York.

images of massive earthworks in India, Australia and North America."[60] Earthworks artist Robert Morris visited the Nazca Lines and on his return wrote: "Everyone I spoke to in Peru advised me to ... see the lines from the air. ... Aerial photography returns us to our expected viewpoint. Looking down, the earth becomes a wall at 90 degrees in our vision."[61] Seeing from above creates new canvases. Whether land art is emergent or disruptive is a question I will leave to others. Whatever the case, new technologies create new politics in new places.

III IMPLICATIONS

6 PROTEST TECH: HOW NEW TOOLS GET ADOPTED, AND WHY THEY SPREAD

Animal rights activist Steve Hindi films hunters. The logic is simple: most people would disapprove of violence toward animals if they were to see it for themselves.

That's why video is important to Steve and his organization, Shark Online (SHARK). The mostly volunteer organization has been documenting animal abuse through film for two decades. The group makes an annual pilgrimage to document Oklahoma Senator James Inhofe's pigeon shoot. "We used to just pull up on the road and set up our cameras," explained technician Mike Kobliska, "but then they pulled back, beyond our line of site. So we bought a small helicopter." Eventually, the team replaced the RC helicopters with custom-built octocopters that allow them to fly directly over the shoot itself. From the air, they document the release and slaughter of the birds and the subsequent bulldozing of the carcasses (I have watched the footage; it's not for the faint of heart). Directly over a pigeon shoot is a great place to film a pigeon shoot—but it is also a great place to get shot by pigeon shooters. That's exactly what eventually happened: the hunters took aim and shot down the group's octocopter, demonstrating the utility of the kinetic interdiction approach described in the previous chapter.

Adopting new technology is not a straightforward affair. The organization also targeted the Philadelphia Gun Club (PGC), a who's who of the wealthy stretching along the Main Line corridor that traces a band of wealth out of New York City and into the Pennsylvania countryside. That group *also* holds a pigeon shoot. Twice a month. When members arrive, SHARK is there, moored in the Delaware River in a boat with their camera equipment. Since two can play at that game, the PGC erected a wall, effectively blocking activists' view of the proceedings. As the wall was going up, however, drone prices were coming down. Hindi and his colleagues sold their boat and continued to monitor the shoot by drone.

Despite resistance, SHARK is going strong. The reasons for this are three-fold. First, new technologies fit solidly within their theory of change. Kobliska explained to me: "People need to see something. You can tell them that it's bad, write thousands of words, but that doesn't have the impact of seeing it happen to animals." If you need video, and the actor your campaign is targeting builds a wall, then you get over it somehow. "We started out with a helicopter," Kobliska reminds me. This flexibility points to a second reason SHARK's work continues. They have demonstrated disruptive creativity in their back-and-forth interactions with those they target. This is true for the Philadelphia Gun Club struggle, but can also be seen in a further example. The group is using remote cameras controlled by Raspberry Pi and Arduino microprocessors. These devices are used to track state wildlife officials in Illinois as they flush and kill deer. When the officials learned of the cameras, they began panning the woods with their flashlights in order to identify and immobilize the devices before commencing the hunt. SHARK responded by devising light-responsive lens caps that would swing down, cover the glass lens, and thereby eliminate glare and reduce the likelihood that the camera would be discovered and dismantled. Kobliska chuckled, telling me "now they've switched to infra-red" to identify the cameras. "It's an arms race," he said: "We find one way, then they bump up their end." Meanwhile, the Pennsylvania Gun Club has brought a lawsuit against SHARK. All signs point to the club's demise, as unfavorable press coverage has led to the suspension of the event.[1]

In this chapter, I argue that the experimental adoption of new technologies may be predicated on their visibility, accessibility, affordability, usefulness, and appropriateness. This case also illustrates the fact that technology can be used in numerous ways, but change-oriented actors, and especially social movements, tend to adopt them for three primary reasons: to gain clarity on a situation, to communicate to the public, or to raise the cost of the status quo. As noted in the first chapter, an important line of scholarship focuses on the role social media plays in advocacy work. This communication capacity is critical if advocacy efforts are to gain attention, influence the public, and change policy. It is therefore important to focus on the tools that help movements communicate with their followers, bystanders, the media, and those in power. Yet movements also use technology to gain clarity about the issues at hand, well prior to and possibly instead of public action. Likewise, movements are keenly focused on rendering business-as-usual practices unsustainable—that's the logic behind boycotts, sit-ins, and investigative journalism. If the maintenance of old patterns of behavior becomes too expensive—in terms of lost revenue, votes, or public support—the thinking goes, then the targeted behavior will change. To push beyond a narrow focus on communication, it is important to emphasize the wide range of tools and technologies that have a real impact on the process of understanding key issues and communicating critical messages. I have written this final chapter in an attempt to emphasize the non-digital technologies at work in fairly traditional advocacy efforts. The larger argument is that there is much to see if we think broadly about tools in use, rather than narrowly about new digital technologies. Artifacts are important for gathering data, making decisions, disseminating information, raising costs, and so forth.

In other words, tools needn't be new or digital in order to be disruptive.

MOVEMENT ARTIFACTS

A world of technology exists before and beyond the tools and technologies that social media represent. There is more to human rights advocacy than awareness-raising, and more to social movements than getting people onto the streets and aligning their demands to what's feasible. Technologies are used to communicate, but they are also used to gather information. Technologies are used to connect people, but they are also used to change the cost-benefit calculus of incumbent lawmakers, elites, or rights violators. In the remainder of this chapter, I apply this book's approach to technology as "tools in use" to a number of cases beyond drones, satellites, kites, and balloons.

Change-oriented social actors craft communication opportunities out of the materials they have at hand. Social-movement scholarship on tactics suggests that individuals, institutions, and initiatives promoting change are often pressured to boil their assessment of a situation down to a slogan, banner, placard, petition, chant, or tweet that can frame the issue for a broad audience.[2] Lengthier promotions come in the form of press releases, email campaigns, and websites. These proclamations link to reports, raw data dumps, curated datasets, documentary films, archives, installations, and performances. One need not say more about this phase, as it is easily the most broadly covered in the literature on social movement technology.

The story of what happens to this raw material is increasingly focused on the spread of particular messages through social networks, often facilitated by social media and then amplified by computational propaganda. This is a terribly important phase of the process, but its operation has already been well told by many others, especially Phil Howard, Jennifer Earl and Katrina Kimport, Lance Bennett and Alexandra Segerberg, Bruce Bimber, Andrew Flanagin and Cynthia Stohl, Andrew Chadwick, Taylor Owen, Evengy Morozov, and Zeynep Tufekci.[3] Readers interested in compelling accounts of the way political actors use technology to generate the pivotal messages that impact public life should start with their work, which I selectively review in an the theoretical afterword.

A sustained focus on the importance of social media might lead one to believe that tweets, memes, and viral videos are social movements' primary mode of technological engagement. Anyone who has made it this far in this book is under no such illusion. A focus on the *stuff* of advocacy communication emphasizes instead the tools and technologies that are used to draw attention to the data gathered via smartphone, data mining, interviews, or geospatial affordances. Movements produce a copious quantity of everyday stuff about which there is little mystery: pamphlets, websites, reports, datasets, press releases, documentary films, and photographic exhibitions. I will not inventory them here, as the point is by now clear: taking technology seriously requires taking communication infrastructure and ecosystems seriously. If we care about social media, we must care about the tools and systems that make it possible.

Capturing data—Critical geographers long ago noted the implications of enhanced data capture. In the inaugural 1969 issue of the pioneering journal of critical human geography, *Antipode*, Jeremy Anderson mused: "May I look at you? Listen to you? Smell you? Feel?... May I overfly you? May I remotely sense you?"[4] Anderson's interest in the moral problems of remote-sensing technology was in direct response to the "development of miniature cameras, telephoto lenses, highly sensitive microphones and miniature tape recorders... [and] air- or satellite-borne remote sensors." A flash-forward to the present finds geographers at work with every one of these tools, and many more to boot. They are not alone, as arrays of sensors herald a computational planet stitched together by sensors in the billions.[5] Camera-equipped devices are one of many visible data-capture technologies within civil society's technological repertoire. Advocacy groups also rely on imagery captured by an array of supportive technologies, such as SHARK's use of Raspberry Pi and Arduino processors, light-detection sensors, and servo-connected lens covers. Critical geographers also debate the use of tools—in some cases for decades.

While the question *may I overfly you* is alive and well in political and cultural debates about drones, the practical answer from many quarters has been simple: yes. At the University of Nottingham's Right Lab, my colleagues rely on satellite imagery to help better estimate the scope of the brick kiln industry and to better inform advocacy efforts on the ground.[6] While these early efforts relied on crowdsourcing, the goal was to help process massive amounts of data rather than raise public awareness. The fact that the project gained popular attention was an unintended byproduct, not our original objective. A range of technologies, old and new, are important for gathering the kind of data nonprofits need to make important decisions, scientists rely on to conduct research, and policy makers use to inform public policy.

Transmitting data—In order to gain clarity about the world in which they operate, change agents must circulate the data they have secured. Of course, they may choose not to, instead securing it against leaks—yet, nevertheless, the data has a life in filing cabinets or on servers. Sometimes these materials have a social life as they circulate through the body politic, as seen in the case of image-making and sharing online and by social media. Often overlooked in this process, however, are the infrastructure, objects, and processes that ease and ensure the movement of data, whether digital or analog. Increased attention to infrastructure has emphasized the importance of the undersea trunk lines that allow the Internet to span the globe[7] and as venture philanthropists launch gliders to provide Internet connections to underserved areas.[8] An emphasis on the physicality of these ligaments is also a key vulnerability: networks are difficult to establish (as seen in a recent and marvelous history of the Soviet Internet),[9] easy to wall off (as evidenced by China's Great Firewall, the semi-permeable membrane that insulates the country from the world),[10] and relatively easy to simply unplug (as with Egypt's disconnection of the Internet during the Arab Uprising).[11] We must remember that the Internet can be located, has thingness, and is comprised of material objects maintained by human agents. This is evident in the case of algorithms, virtual private network services, proxy servers, and the popularity of the "dark web." Existing at the intersection of digital spaces and physical servers, these ungoverned processes and spaces possess significant political potential[12]—some have evolved into micro civil spheres.[13] Equally important are those places where technical and social processes overlap, as in the case of crowdsourcing.

Efforts to describe the relationship between technology and society must account for the tools and technologies that facilitate the movement, distribution, or dispersal of change-related data. In some contexts, this can be conceptualized as *infrastructure*, including the microchips, digital devices, modems, cables, routers, switches, branch lines, servers, trunk lines, service providers, and satellite feeds that comprise "the Internet." In other cases, it might be thought of as material *resources*, including land, offices, automobiles, accounting procedures, and fundraising tools, including bake sales, charity auctions, and recurring contributions. Taken together, these infrastructural features are the technical delivery systems for the stuff that change agents traffic in: symbols, money, memes, messages, news bulletins, press releases, declarations, and so on.

They all flow over something.

CNN's early and consistent coverage of the 1989 protests in Tiananmen Square is a striking example of the important role played by communication infrastructure and networks. The anti-government protests that brought Chinese students to the street in 1989 also attracted Western media outlets. CNN's team—40 people in total—were already in China in order to cover a visit by Mikhail Gorbachev and a conference held by the Asian Development Bank. The video feed coming from CNN's cameras did something revolutionary: it went straight to space, ricocheted off a satellite in geosynchronous orbit, fed directly into CNN's programming, and was piped to televisions in American homes and those Chinese hotels that catered to the international community.[14] Chinese officials immediately recognized that a satellite signal had bypassed their sensors and censors in new ways.

They cut the feed.

What makes the use of satellites so compelling in the case of Tiananmen Square is their pivotal invisibility as a conduit for instantaneous and worldwide press coverage. While militaries had this ability decades prior, more democratic control of the means of distribution by an institutional actor like CNN posed a threat to an oppressive regime, which responded with both military and technological force.[15]

My first job out of grad school was working for a human rights advocacy group. Disturbed by the post-9/11 rise in the use of mercenaries and forced labor—a case of hegemony outsourcing the dirty and dangerous job of both tearing down and building up—we launched a campaign called WarSlavery. The project focused on reports that the new American embassy in Baghdad was being built with forced labor. Contemporary slavery was embedded in the network of contractors tasked by the Department of State and Department of Defense with the rebuilding of America's premier diplomatic symbol. Some city on a hill, we thought. As I worked with a team to roll out the campaign, the question quickly emerged—how should we coordinate a global advocacy network on such a sensitive topic?

At that time our organization used an online team management environment called Basecamp. Our team debated the opportunities and threats of any similarly cloud-based approach, including the Google suite of services many advocacy groups have turned to in order to save money. Public opinion was still generally in favor of the war,[16] and we were pretty sure the Federal government was having its way with any data it could get its hands on, legally or not. While time has chastised the war's proponents and shown the wisdom of those fearful of PATRIOT Act–inspired overreach, this future was not obvious in 2004 as we started the campaign. In retrospect, we were debating how to harden our campaign against the prying eyes of its target at the same time the Republic's security apparatus was implementing large-scale and illegal sweeps of personal data, extralegal extraterritorial renditions, torture, and the murder of American citizens. It is now clear to the most casual of readers that we should have done a better job hardening our backend infrastructure against a government that was hard at work targeting anti-war protestors and gathering as much digital data as possible. The risks inherent in hackable files stored in online servers is demonstrated by an ongoing litany of high-profile leaks.

A final case perhaps rounds out my argument that transmitting data is an important aspect of technopolitics. High-profile leaks of classified government data—Edward Snowden, Julian Assange, and Chelsea Manning being the best known—appear to be digital affairs, but also reveal the extent to which overlooked aspects of technology actually matter. Data's media, in a manner of speaking, has gone from reams of paper and bulky mainframes to linked networks and streams of data. Isolating digital materials in an offline location doesn't preclude one from badging a rewritable compact disk with a Lady Gaga cover, sliding it into the non-networked drive, and walking away with the goods. The importance of this physical medium—bits and bytes rather than pages and pages—deserves our attention. A typical compact disc can hold around 700MB of data, which equates to roughly 175,000 pages of typed data. I doubt you have a shelf of encyclopedias at home, so I'll remind you that the Encyclopedia Britannica runs about 32,000 pages. The entire neighborhood would notice if you walked out the door with it. By comparison, Chelsea Manning's rewritable Lady Gaga disk could hold five encyclopedia sets worth of data. For younger readers, that's the physical equivalence of 11,072 smartphones—not their memory, but their actual bricky bodies. For older readers, the weight of that many printed pages would be nearly equivalent to that of a 1969 VW Beetle—hard to budge between cubicles, and even harder to fit into the elevator.[17]

For readers of all ages, the point should be clear: no digital storage artifact means no Manning, no Snowden, no Wikileaks, no way.[18] Modality matters. Materiality matters. Collective-action efforts generate important data that must be stored and shared somehow. These physical and technological aspects of the story—the how—often take a back seat to the subsequent who-what-when-where-why of the event. The importance of CNN's role in telling the Tiananmen Square story to both the world but also to Chinese citizens is well known. Less dramatic, but equally important, is the technology that allowed the story to be told.

While we know quite a bit about Manning's medium of choice, my efforts to determine which satellite CNN used to transmit Tiananmen Square data were futile. It might have been from IntelSat, or might have been RCA's SatCom 1 or the Statsionar-12. Nobody's really sure. Stories about tools and technologies in which the characters are material bits, bobs, and bots may not make for a ripping read, but they are a critical component of social change efforts worldwide. Technology companies interested in closing the divide between the digital haves and have-nots recognize this fact. Both Facebook and Google are testing high-altitude Internet platforms in the form of UAVs. These gliders and balloons are seen as the ultimate leapfrog technology, skipping entirely the sort of infrastructure that had once been the state's duty to provide and regulate. Only time will tell whether such a move emancipates web traffic from national servers and firewalls or stifles citizens everywhere under throttled mediations and invasive user agreements.

Analysis—What would a materialist account of data analysis look like? By *analysis*, I simply mean the process of meaning making required to translate data into information. Analysis takes many forms, since social actors translate data into information all the time, often without thinking about it. It may be quantitative, in the form of public opinion data, or creating a baseline estimation of brick kilns in India. Or it may instead be qualitative, as with secondary desk reviews, interviews, focus groups, imagery analysis, and so on. More often it is comparative, as individuals and institutions look at variation: over time, between cases, or between varying interpretations of the data. Analysis may also be generative and conceptual, as when organizations come up with compelling language and work to explain sets of relationships or important background factors. Analysis may be causal, in an effort to explain why something happened or the likely consequences of an anticipated action. Analysis may also be legal, as institutions attempt to determine who is legally responsible for a particular action or chain of events.

These analyses are facilitated by increasingly powerful computational devices, but the materiality of this process remains the same: humans and their digital and analog companions, the computer and the legal pad.[19] I have given human examples, because I believe this task largely remains a human endeavor. New technologies are helping to gather new data, and more powerful computers help to crunch that data more efficiently, but the act of saying what it means—of analysis—has generally remained a distinctly human endeavor. This may change under two conditions. The first is in the spread of linear predictive algorithms, which are programmed for action in response to input. At some point in the future, a human rights violation may be determined to have occurred because the data suggests this is the only logical outcome from the empirical evidence. This is a possibility under the condition of human-programmed algorithms. A second possibility lies in the emergence of artificial intelligence, in which a sentient nonhuman life agent performs the analysis.[20] Should this possibility become a reality, there is no way to determine whether subsequent decisions by any superintelligence would favor the kind of human rights norms developed by Westerners in late capitalism, or prefer an alternate human-devised system of ethics, or would set off to instead develop an altogether new post-human system for guiding action.

More tools gathering more data being passed around at greater speeds calls for more sophisticated tools for analysis. Expanding networks of sensors only accelerate this process. An initial wave of enthusiasm for the promises of "big data" has broken on the shoals of computational and creative limitations. What are the right tools in the search for data, and for information within those data? What should we be looking for? How will we know when we've found it? Over the next decade, many of these questions will be answered. This same decade will also see the emergence of new questions, as the number of users, devices, and sensors multiplies exponentially. Jennifer Gabrys anticipates a computational planet,[21] and Gina Neff and Dawn Nafus envision the computational self[22]—earth and body shot through with sensors, pulsing data onward by the quintillion.

The Institute for Conservation Research at the San Diego Zoo,[23] for example, draws on a growing range of sensors, all of which generate a significant amount of raw data. David O'Connor, a researcher at the institute, explained to me the data implications for a single pilot conservation project. The institute and its in-country collaborators installed wildlife trap cameras at two conservancies in northern Kenya: the 226 square kilometer Loisaba Conservancy and the 3,940 square kilometer Namunyak Conservancy. The project's motion-sensor–equipped cameras are designed to inventory occupancy and to track the movement of giraffe and leopards, while also capturing photos of all wildlife species to get a sense of biological richness.

When it's all said and done, the group's 120 cameras generate millions of images each month. For instance, during the pilot study, with just 30 cameras at Loisaba, over 350,000 photos were taken. The cameras are placed out in a grid pattern, but since there are not enough cameras to put out on all points of the grid at once, cameras are rotated through the area. The grid covers 100 percent of Loisaba, but only 25 percent of the grid is equipped with cameras at any point in time. By comparison, the grid for Namunyak only samples about 6 percent of the total conservancy area (with 3 percent equipped with cameras at any point in time).

A straightforward project with a limited scope (two patterned animals) will generate about 12 million images a year. Setting up a similar system to track these two species over the range of reticulated giraffe in Kenya—they roam over the entire area of the Northern Rangelands Trust Community Conservancies (44,000 square kilometers) as well as an additional protected area of just over 10,000 square kilometers—would generate over 43 million images in just one month if camera traps were placed in a 5-square-kilometer grid pattern (about 10,843 camera traps). Implementing such a program is complicated, requiring new data capture tools (cameras, GIS-transmitting satellite collars) and new infrastructure for transmitting this raw data (satellite, Wi-Fi). A program that generates 516 million images per year would require the kind of data-analysis tools nonprofit actors like the San Diego Zoo tend not to have. The massive computational horsepower and storage volume needed is simply out of reach of all but the most well-resourced institutions.

The central problem in data analysis of any sort involves separating signal from noise. Many of the images made by the Institute capture moving things that are neither giraffe nor leopard. The sensitive devices are as likely to capture a leaf blowing in the wind as they are local pastoralists taking a whack at the cameras out of a fear that their herds of cattle are being monitored for illegal grazing. There are several methods for isolating what's important in the resulting data. Analysis can be done by hand, whether through a large staffing expenditure, a long-suffering volunteer, an open crowdsourcing campaign like Zooniverse, or the use of a crowdsourced marketplace like Amazon's Mechanical Turk. Alternately, the process could be automated, though this step is complicated and generates false positives. Once the desired subset of images has been identified, the task gets much easier. The Institute's project tracks the animal's movements by cross-referencing the camera trap images with a database of known animals. Cross-referencing is automated. Giraffe and leopard coat patterns are unique. It's a perfect task for an algorithm.

New tools and better infrastructure create both opportunities as well as challenges for intergovernmental, nonprofit, community-based, and social movement organizations. Individuals as well as organizations have an opportunity to better understand complexities within their areas of interest. The challenges are equally clear. New opportunities often require new resources, and resources aren't free, as every movement scholar from Robert Michels onward can attest.[24] Michels famously coined the phrase *iron law of oligarchy* to describe what happens when people organize for efficacy.[25] At the end of the day, organization creates divisions of labor that require expertise, and the entire affair produces inequality—or, in Darcy Leach's pithy summation:[26]

> Bureaucracy happens.
> If bureaucracy happens, power rises.
> Power corrupts.

That was as true for Michels' Italian anarcho-syndicalists comrades organizing in 1911 as it is today: it takes money to make these things happen. Starting from scratch is costly. Scaling up is expensive. Sustainability may be a disappearing point on the horizon. Big resources make a big difference, and as a result organizations can do more of the stuff they set out to do. Critics like Michels, and more recently Francis Fox Piven and Richard Cloward, rightly note that resources may have a moderating effect that dooms radical efforts of any stripe.[27] Resources are sites of contention.

It should be no surprise that there is a back-and-forth struggle over the means of cultural production. Sometimes the people have the upper hand in the struggle for hearts and minds, as with the adoption and diffusion of solidaristic hashtags (#iranelection, #blacklivesmatter, #umbrellarevolution, #metoo), and at other times the regime has the upper hand, as Egypt throws the OFF switch for the Internet or when Russia and Facebook flip the ON switch for fake news. While debate rages over whether hashtag activism is *real* activism, the important thing to note is that regimes consider Internet-facilitated channels of communication—IRC, Facebook, WhatsApp, Telegram, WeChat, Signal, Twitter, Instagram—to have sufficient organizing, framing, informational, and mobilization potential to merit hacking, hijacking, or shutting down. These programs combine many of the factors under consideration in the pages that follow: input (camera-equipped smartphones were pivotal to the emergence of the Black Lives Matter movement), throughput (via Internet service providers, telecoms, apps, and social media platforms), analysis (in the form of comments and hashtags), and the output found in the next section.

Change agents and incumbents rely on a range of tools and technology that are broader than we usually realize. Materiality matters, and techniques for facilitating or inhibiting social change are all around us. Although digital accomplishments are important, the analog matters as well, as a historic case nicely demonstrates.

MOVEMENT ARTIFACTS: ANTI-SLAVERY PETITIONS

Anti-slavery efforts to petition the British Parliament began in the late eighteenth century and stretched on until the 1830s. The first round of petitions emerged at the height of the transatlantic slave trade. Britain's economy relied heavily on trade-related commerce, yet abolitionist fervor rocked the country. A narrative of national confession spread as religious leaders framed slavery as a national sin, with immediate abolition being the only righteous sacrament.[28] When a group of radical Quakers formed the Anti-Slavery Society, their strategy was twofold. They would turn public opinion against the practice, and use that inertia to pressure Parliament to end the slave trade. These strategic objectives came together in a singular movement tactic: the mass petition.

Between 1787 and 1833, hundreds of petitions were compiled in England, Scotland, and the United States. These were politically symbolic actions and objects. For example, the "Great Women's Petition" of 1833 gathered signatures on individual pieces of parchment. These parchments were then stitched together to comprise massive single rolls: one roll of 179,000 signatures to the House of Lords and another of 187,000 to the House of Commons. Seymour Drescher noted that the "petition to the lower House was nearly half-a-mile long."[29] In the late eighteenth century, 519 petitions were brought before Parliament, nearing a total of 400,000 signatures.[30] A celebrated historian of the era, Seymour Drescher, noted that at 187,000 signatures, the petition destined for the House of Commons was a "huge coil" that required four MPs to carry and would have unrolled to a length of nearly half a mile.[31]

While these numbers themselves are impressive—one in five adult males signed British petitions in 1814 and 1833,[32] and perhaps as many as half signed the 1792 petition in Manchester[33]—what I want to highlight is the petition roll as a physical representation of the will of the people. It was the act of carrying it into Parliament—where it landed with a metaphorical and literal thud—that resonated. The petition was but one of many key movement tactics pioneered and popularized in the era. A partial list includes the novel adoption of banners, economic sanctions and boycotts, lawsuits, legislative challenges, pamphlets, safe houses and sanctuary, slogans, petitions, and physical violence.[34] Such adoption did not occur in isolation, but relied instead on a critical shift in the broader political economy, as the "concentration of capital and the related proletarianization of the British workforce altered the interest of workers and employers"[35] alongside the availability of raw materials for protest. These tactics are both physical and symbolic, and for their efficacy rely almost entirely on a constellation of associations, assemblies, publics, public opinion, and the press. With the exception of safe houses, each is meant to be seen, heard, and, in some cases, felt—both physically and economically. The iconic abolitionist image "Am I Not a Man and a Brother" was reproduced on material objects, including hairpins, purses, jewelry, and snuff boxes. These were simultaneously consumer purchases as well as political signals.

The materiality of this advocacy artifact gave it a public and political life. Likewise, abolitionists' signatures made a political difference. But that was magnified by its material form, a physical representation of the will of the people that rested heavily on the shoulders of the body politic.

MOVEMENT ARTIFACTS: INTIMATE PARTNER VIOLENCE

The importance of a materialist read of advocacy efforts can easily be seen in digital forms of engagement as well. While we tend to think of mobile-phone-mediated engagement in terms of social media, Project Concern International, a nonprofit working on issues of health and development worldwide, joined up with the public relations firm Ogilvy to craft a social engagement campaign on domestic violence in South Africa. Work in South Africa is critical, as the country has some of the world's highest rates of coercive sex, rape, and intimate partner violence.[36]

The group's response to the problem was unique; they created an awareness campaign that included the installation of two five-story billboards along major traffic routes in Cape Town and Durban. On one of the billboards, for example, was a woman's face—unmistakable in scale but otherwise unremarkable—with the tagline "What is keeping violence against women alive today?" These images were simultaneously reprinted in local newspapers. After two days, the billboard's surface was altered to show the same woman, but with visible signs of abuse (figure 6.1).

This process was repeated every few days as the signs of abuse grew more severe and as a new tagline declared "if you believe she deserves this, it'll just get worse." As the week passed, and as signs of abuse increased, the text changed: "Change the beliefs that keep violence against women alive. SMS 'stop' to 38797,

Figure 6.1
Western Cape Network on Violence Against Women (used by permission).

and it will get better." At this point, and as text messages came in from passersby, the image sequence was reversed, the woman's face began to clear, and the text shifted to a new message: "Change beliefs ... and it will get better." Rallies held beneath the structures in both Durban and Cape Town, as well as the campaign's own data, suggested more than 20,000 people had directly participated in the event, both in person and via text messaging. While the presence of people and digital engagement matters, it is the physicality of the output that deserves notice.

The campaign deserves mentioning not because it featured a billboard—a staple of public awareness campaigns. Neither do its merits lie in the shocking content. This is, after all, the longstanding commitment to *witness* that motivates groups like SHARK to visually document violence against animals and Alice Seely Harris to document slavery's cruelty in the Congo. What sets this campaign apart is the intersection between text messaging—a newer but broadly distributed and affordable technology on the user end—alongside a billboard, an expensive but traditionally analog media platform. This intervention struck a delicate balance between organizational and popular social resources. It did so in such a way that an object—a billboard—was transformed through the digital intervention of physically proximate bystanders.

This is hybridity at its best.[37]

Many other examples will come to mind for readers familiar with change-oriented advocacy efforts. My goal is to direct attention to the importance of materiality. I could just as easily have pointed to art installations, annual reports, academic investigations, or protestors' barricades. Across these cases, I am arguing that new data-gathering tools lead to new ways of seeing, which in turn lead to new ways of showing. This observation is not limited to drones, geospatial affordances, or even digital technologies. Analog and decidedly old-school processes (parchment petitions) and mashups between communication and advertising tools (mobile phones and billboards) point to the flexibility of this approach to movement technology used by change-oriented social actors like nongovernmental organizations and social movement groups.

Advocacy efforts rely on public demonstrations of worthiness, unity, numbers, and commitment.[38] It is crucial that a group's light not be hidden under a bushel, in a manner of speaking. Getting the light out from under the bushel and onto the street is often a physical activity with material components. Numbers, especially, are critical in this process. Protest size matters because that's how worthiness, unity, and commitments are made public.[39] Of course, other metrics matter as well. Dollars given and votes cast are critical measures of public support for important issues, but they appear on balance sheets and disappear into halls of power. A focus on output is sufficiently broad as to capture spontaneous crowds, confusing budgets, and forgotten votes as well as stuff with lasting materiality. The petition that landed with a thump in the House of Commons matters.[40] Its physicality mattered, communicating the importance of words written elsewhere. None of this is to imply a linear relationship between the intention of the material's creators and their ultimate utility. Sociologist Terence McDonnell has compellingly demonstrated the ways in which cultural entropy reclaims the material objects of high-priority efforts at social change.[41] Advocacy groups may document the success of such an installation in the moments immediately following its deployment, but there is another story to be told in the longer life that material goes on to have in other spaces and in other hands, and for new reasons altogether.

MOVEMENT ARTIFACTS: RAISING COSTS

Movements to end slavery or intimate partner violence are driven by people with a simple theory of change, and one clear goal: transform attitudes and behavior. Changing both hearts and minds is best, but we're often willing to settle for changed behavior. That's why political and social change efforts focus on "awareness-raising" and on raising the cost of compliance beyond what the incumbent is willing or able to bear.[42] Much is known about awareness-raising and political communication, but social movements also work hard to make the status quo too expensive for the folks they're targeting.

Cost-raising relies on oft-overlooked tools.

Here I am thinking of material objects that make the status quo unsustainable for those it has traditionally benefited. This is in contrast to reputational costs, which often take the form of the bad publicity that hurts market valuation or electability. Brayden King, Sarah Soule, and their colleagues have done wonderful work documenting corporate sensitivity to public opinion, itself a proxy for the support publicly traded firms enjoy in the stock market.[43] A high-profile story about corporate malfeasance, whether in the form of corruption or abuse, can lead to devastating losses and mass sell-offs. These efforts materially hurt publicly traded targets of advocacy efforts, but they are not the focus of my attention here. I am more interested in the tools and technologies that are used to raise the cost of the status quo.

The most dramatic form of cost-raising involves the violent use of weapons. Protestors and insurgents willing to use physical violence run the risk of alienating the general public, to be sure, but they also significantly raise costs for those they target. These costs are reputational, as the public wonders if authorities can continue to provide for public safety. These costs are also material, in that they require the deployment of additional security measures, including police paid overtime and the hiring of specialists called up from highly trained and more expensive contingents. These security forces must also be equipped with specialized and costly offensive and defensive gear. For their part, violent protestors bring to bear a range of tools. These may be offensive, as with baseball bats, rocks, paving stones, Molotov cocktails, foodstuffs, shoes, tear gas canisters (returned to the police), urine, glass bottles, firecrackers, smoke bombs, eggs, garbage, metal barricades, burning bottles, flags and t-shirts, contaminated water, chairs, chains, paint, and the like.

Nonviolent street protests and marches extract many of these same costs, especially when states and powerful corporations draw on riot police, whether or not protestors have used violence. Nonviolent sit-ins and die-ins draw on the symbolic material of the human body to block access to public and private space. Prostrate and interlinked bodies are difficult to move, and they disrupt normal flows of traffic on roads or the passages in and out of buildings. Student protestors who have linked arms around a university building, for example, impede the institution's ability to perform normally. Disruptive events create an incredible sense of solidarity and are high-visibility opportunities for the movement as well as threats to a university's leadership and public image. Visibility, and these costs, are earned through a particular tool: the body of the protestor.

Barricades are another disruptive and cost-raising tool.[44] Like the prostrate or arm-linked corpus of protestors, barricades immobilize the flow of traffic, challenging business as usual and forcing attention from authorities and the public. These costs were too high for Napoleon, who famously enjoined Georges-Eugène Haussmann to widen the avenues of Paris. This would reduce the crowding that led to "misery, pestilence and sickness," claimed Victor Considerant, the social reformer.[45] Of course, it also reduced the likelihood that urban insurgents could block-ade their narrow streets as they had done in the past. Barricades are comprised of the material at hand, whether a prised paving stone or park benches, dumpsters, earth, repurposed police bar-ricades, sandbags, vehicles (police, taxi, private, bicycles, busses, streetcars), tires, trees, furniture, and anything else within reach, including general detritus. Mark Traugott's delightful and pains-taking work on the Parisian barricades demonstrates the interplay between the broad barricade repertoire and its individual and evolving instantiations. In keeping with the general argument that particular repertoires are the product of their times, we can see that Parisian barricades in 1877 included objects that were not available 30 years prior, namely streetcars, tires, and urinals.[46] The strategy remained stable over a significant period of three decades, but its tactical substrate, if you will, changed with the times.

Strikes, whether peaceful or violent, raise costs for incum-bents because they withhold a critical material input from the production process: bodily labor. As with sit-ins and die-ins, the corporal body is the material object that makes a difference. Where sit-ins rely on the presence of bodies at the wrong place and the wrong time—laying in front of a university president's office during business hours, for example—strikes rely on the absence of bodies at the moment they are needed most: as units of force in the mode of production. Strikes may be combined with barricades to keep owners and managers out of production facilities, and may be combined with weapons, as is the case when strikers clash with police or private security forces tasked with strike breaking. Sabotage and machine-breaking are specialized forms of protest violence directed against physical means and modes of production, and may be combined with strikes or conducted separately.

On a similar note, I am often reminded of James Scott's anecdote in *The Art of Not Being Governed*: when enemy armies come to plunder the harvests of peasants, he writes, they are "powerless against the lowly potato."[47] Since they grow underground, tubers can be hidden much more easily than crops like wheat and rice. Such *escape agriculture*, Scott argued, has facilitated the autonomy and mobility of the powerless in places as diverse as North Carolina and Southeast Asia.

Hacking also raises costs for incumbents, as it undermines confidence in their ability to keep crucial data safe while also raising provocative new questions about the information found in the data itself. High-profile data dumps from WikiLeaks triggered a wave of actions from the American government intended to reduce the revelations' impact on its military and commercial interests. Here the tools of the trade are both digital as well as analog. Some combination of sophisticated programing skills, direct access to classified files, access to illicit troves of passwords and security keys, and a certain degree of intelligence about human behavior form the ingredient list for most high-profile leads. But every one of them requires a script or algorithm running on a computer, a USB drive smuggled into a secure site, or a rewritable compact disk disguised to look like a Lady Gaga CD: every hack requires material tools of some sort.

MOVEMENT ARTIFACTS: VISIBLE, ACCESSIBLE, AFFORDABLE, USEFUL, AND APPROPRIATE

In this volume, I have argued a number of things: that drones, kites, balloons, and satellites (i.e., geospatial affordances) represent an important new tool for crucial civil society actors like social movements, nonprofits, and intergovernmental organizations; that geospatial affordances create new public spheres; and that the use of new technologies may be emergent and disruptive. In particular, I have highlighted the ways drones, balloons, kites, and satellites can be used for the public good. What I have not done is offer an opinion on what shapes adoption of particular technologies; nor have I predicted the direction of future developments. In this section, I introduce a number of brief hypotheses that may guide our thinking about whether and when change-oriented actors—and here again I am focusing my attention on social movements—adopt a particular technology.

Tools, like tactics, are a product of their time, and they emerge and diffuse based on their real and perceived applicability. How exactly things get used, media scholars Gina Neff and Peter Nagy argue, is the result of the interplay between the perceptions, attitudes, and expectations of users, the materiality and functioning of the technologies themselves, and the perceptions and intentions of an artifact's designer.[48] Interpretive flexibility—the idea that things can get read in a number of ways—means there is no guarantee a particular tool will get used in a particular way by a particular actor, at a particular point in time, *just because it's there*.[49] Some things are simply "more difficult to position in mind, purchase and use, require more support from social contacts, and are only meaningful in selective contexts,"[50] and as a result no two people are guaranteed to see a technology's utility in exactly the same way.

I believe five key factors shape adoption. Basic *visibility* is a necessary but insufficient condition. In order to enter an individual or institutional repertoire, and perhaps to even be considered or experimented with, a tool or technological solution must also be considered *appropriate, accessible, affordable*, and *useful*.[51]

Visible—At the most basic level, a tool must be visible. It must be seen and considered for use. This argument is clearest in scholarship on affordances, since it is the "relationship between the properties of an object and the capabilities of the agent" that determine use, and since the "presence of an affordance is *jointly* determined by the qualities of the object and the abilities of the agent."[52] Presence is established through the senses—we cannot use a thing if we cannot perceive it. My use of the word *visible* is meant to signal perceptibility or conceptual ascertainably, rather than "seen by the eye." No surprises here: if an actor doesn't recognize a thing as having use, the thing is not of use to the actor.

Accessible—Social movements adopt tactics and technologies that are within their real or perceived "toolkit" or are available in their broader milieu. Appropriate institutional approaches are often based on and embedded within broader organizational, social, and cultural norms,[53] and there is no reason this logic does not extend to technologies. It is reasonable to hypothesize that organizations will choose to adopt technologies only if they are legible to decision makers and can be incorporated into their existing tactical repertoire in a way that resonates with key stakeholders.

Affordable—Change-oriented actors adopt tactics and technologies that they can afford. Social movements often mobilize around areas of injustice in which resources or recognition are not fairly distributed. Though there are important exceptions to this broad statement, the reality is that change-oriented actors often face resource constraints. As a result, the presence or absence of key economic, institutional, and personnel inputs can spell the difference between mobilization and inaction and between success and failure.[54] Perhaps better-funded organizations are more likely to consider or use new technologies. This is important to emphasize. The democratized surveillance described in this volume is predicated on access to key resources, and resources are unevenly distributed between sectors. Businesses, governments, and large institutions have a wider range of resources than marginal institutions and nonprofit actors. Resources are also unevenly distributed across regions, with urban areas often having more resources of certain types than rural areas, and with organizations in the Global North having more of certain kinds of resources than do southern organizations. The concept of *affordability* is therefore relative to the particular context change-oriented actors operate within.[55]

Useful—Social movements adopt tactics and technologies that they believe will help their cause. Over the last century and a half, cameras have been a critical tool in the advocate's toolkit.[56] More recently, satellite technology provides an excellent opportunity to capture forensic evidence from a new perspective.[57] On the street, ersatz t-shirts protect against tear gas, while online hashtags are increasingly able to draw attention to issues and, in the case of #blacklivesmatter, #umbrellarevolution, #metoo, and #timesup, frame an entire movement. These things are used because they are considered practical and useful to the task at hand, whatever that task may be. In order to pass into a repertoire, this use must be relatively widespread and constant. After all, if people did not "engage in this continuing activity of material and social production, the human world would literally fall apart."[58]

Appropriate—Social movements adopt tactics and technologies that they feel will support core movement goals. However, movements are also fundamentally exposed to and situated within meaning-laden cultural contexts, and as a result movement actors concerned about public opinion will refrain from tactics that alienate the general public.[59] It is the logic of appropriateness that links directly to the notion of disruption used throughout this volume. A technology is disruptive when it lies significantly beyond the political or social status quo. Since the public generally has a negative opinion of violence, for example, it is reasonable to hypothesize that most advocacy groups will refrain from its use whenever possible.

The notion of *appropriateness*, like that of disruption, is highly variable and context-specific. Assessments of appropriateness are made based on the technologies' fit in light of other commitments, histories, and logics.[60] In other words, the diffusion of innovative tools and ideas is fundamentally a social process.[61] This logic can be seen in major human rights groups' ambivalence toward the use of the kinds of UAVs discussed in this volume. Human Rights Watch, Amnesty International, and the Open Society Foundation have invested heavily in a normative critique of the United States' drone-based "targeted killing" campaigns. I believe the inertia of this campaign acts as a drag on their acceptance of smaller-platform devices with a wider range of functions. Movement struggles create their own histories, as "participants remember what happened before and plan accordingly." History casts long shadows, especially since previous challenges lead to new arrangements that themselves become the status quo.[62]

A second form of this logic can be seen in the tension between the ACLU's condemnation of police surveillance by drone and the possibility that they might be useful in documenting police violence. Movements may tinker with new tools and tactics, but they do so in small ways, "at the edge of well-established actions."[63] A third example of the logic of appropriateness lies in organizational assessments of privacy issues prior to the introduction of drone technology. Institutions mobilized on issues of data protection and privacy are less likely to view small-scale drone technology favorably.

Each of these measures of appropriateness is empirically measurable, but none are required. The fear of missing out (or FOMO), institutional isomorphism (imitation is the sincerest form of flattery), and homophily (birds of a feather flock together) guarantee that some technologies and techniques may be adopted without extensive reflection.[64] Furthermore, each of the five factors suggested above are falsifiable, and thus amenable to future empirical analysis. It may be, however, that some factors are mutually exclusive. What happens if a technology is useful but not appropriate, as in the cases listed previously? And what if important actors consider it to be affordable but not accessible as a tool in the repertoire, as incumbent corporations like Palm, Kodak, and Nokia decided about new digital upstarts in the 1990s? It may be too early to tell whether drone technology will enter institutional repertoires, and there is perhaps good reason to anticipate it will not, but it is not too soon to ask what explains social movement attitudes and behavior toward new technology. It seems reasonable to expect that variation in perceived appropriateness, accessibility, affordability, and usefulness explain the acceptance (attitudes) and adoption (behavior) of particular technologies. Each of these factors emerges from a broader technological moment that mediates decision-making about new ways of seeing and sensing.

CONCLUDING THOUGHTS

This chapter has set out to illustrate what a materialist approach to movement activity might look like, in this way reinforcing the book's broader argument that social change efforts rely on a wide range of affordances. This book has taken a peek at things in use before and beyond the new digital technologies that capture the headlines. It also takes seriously the infrastructure of new digital technology. As I write these words, important questions about social media and democracy demand attention. Pioneering work on political bots and computational propaganda by my colleague Phil Howard at the Oxford Internet Institute raises fundamental questions about the very nature of political communication.[65] I am also writing at a time when the democratic process in both the countries I work in—the United States and the United Kingdom—has begun to resemble a science fiction drama. "The Waldo Moment," an episode of BBC's *Black Mirror*, has its viewers imagine the shift into a post-human political space, in which politics is reduced to a digital parody of television rather than the hard work of governance. The nature and direction of political communication is subject to fresh analysis and the notion of hybridity is more important than ever.[66] Technologies that were generally ignored and overlooked have, since the 2016 election of Donald Trump in the United States, been subject to fresh and intense scrutiny.

The result of this flux is that some of the points I make here appear less radical than they did when I first put pen to paper. Into this dynamic space I have suggested that we take seriously a technology in its infancy. What the future holds is anyone's guess. Thus, what began as a brief set of articles on the experience Tautis and I had using drones and balloons has expanded into a broader range of puzzles that I have done my best to catalog and clarify. In particular, I have argued:

- Technology and tools are simply things in use.
- Tools and technology may be analog or digital, old or new, visible or invisible.
- Tools before and beyond the digital are important.
- Affordances are socially acceptable clusters of tools in use.
- Geospatial affordances are tools for doing things in the air.
- Technologies have, and create, politics.
- The materiality of technological artifacts has social and political implications.
- Technologies create new space.
- Technologies create new political realities in those new spaces.
- Agency lies with human living beings.
- Agency of human living beings is shaped by the world as we find it.
- Nonhuman living beings may develop agency of some sort.

- Disruptive technologies violate norms.
- Emergent technologies do things that most actors could not previously do.
- Military drones like the Predator have an alternative genealogy worth considering.
- New technologies create novel forms of social action.
- New technologies elicit social reaction against technology.
- Thinking about technology as *stuff in use* reveals a wide range of overlooked uses.
- Tools get used to gather, store, and spread information.
- Tools get used to raise costs, build institutions, challenge the status quo, and displace the old guard.

This process has also raised significant questions, some of which I touch on and others I've avoided altogether. It seems impolitic to lay these questions at others' feet, but a good many of them are beyond my area of expertise and represent areas where solid work is being done by others.[67] I end this volume with questions others may take up in future research:

- Why do individuals or institutions adopt particular technologies?
- Is adoption indeed predicated on a tool's visibility, accessibility, affordability, usefulness, and appropriateness?
- What is next for the adoption of geospatial affordances by civil society actors?
- What form will future unmanned aircraft system technology take?
- What variation will emerge in terms of use, regulation, and public opinion?
- What factors might comprise an ethics of public drone use?

- What role will artificial intelligence play in the tools we use to encourage or discourage change?
- What kind of politics do geospatial affordances imply or require, and how might this vary over time and space?

These puzzles lay across an eclectic range of disciplines and subdisciplines. In the final analysis, I must admit a certain trepidation that in drawing so broadly from social movement scholarship, civil society theory, communication and media studies, and science and technology studies, I will have stretched any particular argument too thin, and my contribution might slip through the resulting fissures and cracks. Of course the alternative, which I greatly hope for, is that some combination of attributes will resonate with readers focused on different projects in different spaces, and in so doing spark new ideas.

If I am lucky, some of them will be emergent and disruptive.

The bigger challenge, perhaps, is leveled by Langdon Winner, who suggests that as we make things work, we must also ask: *what kind of world are we making*? This question is often asked once it's too late rather than when it is needed most.

Implicit in Winner's observation is the notion that the invention and adoption of new technology has existential implications and should not be left to the market alone. The post-hoc regulation of technology guarantees that publics and their governments enter into a world that has already been made, and the ability to *choose the battlefield*, as Sun Tsu recommends, has already been lost to another. In this way the regulators, users, and subjects of a new technology are faced with a more limited set of options than they might like.

This is not technological determinism, but instead a basic observation about strategies of maneuver. It is imperative, then, that in the early moments of a technology's life we attend to its technics and technique, but that we also attend to its psychological, economic, political, and social implications.[68] It is fundamental that we ask whether new technologies will expand human freedom and control, or inhibit it, and do our best to discipline technology in a way that enhances open societies.[69]

I hope that is what I have done here.

THEORETICAL AFTERWORD

1 THE TECHNOLOGY OF POLITICS, THE POLITICS OF TECHNOLOGY

In the hope that there are determined readers who would like to take a peek at the academic code, I here highlight some of the scholarship that has guided my thinking as I wrote this book. The result is a bit more than an annotated bibliography and a bit less than a literature review. Casual readers are advised to proceed at their own risk, and academic readers are invited to note where I map onto current debates, and where I go off the rails.

In many ways I feel this book is an intervention that traces a number of themes introduced by Taylor Owen in *Disruptive Power*. There Owen provides a counter-narrative to the oft-told story that new technologies will transform social life for the better. His work emphasizes the power technologies have, not just for connecting advocacy groups locally and internationally, but for the entrenchment and enhancement of state authority. Though many scholars of democracy and civil society were quick to emphasize the role of social media in several major movements—notable recent examples include the Arab Uprising, #blacklivesmatter, and #metoo—Owen's work suggests that a more measured approach is necessary. Recent turmoil—whether in the form of Brexit or unsettling election results in apparently settled democracies—has made clear to everyone that digital politics is a two-edged sword that simultaneously increases civil society's ability to mobilize and enhances the power of anti-democratic actors. Concerns over computational propaganda contrast with earlier hopes for liberation technology[1] and writers like Evengy Morozov have been persistent in directing attention to the enduring importance of classic political considerations, including the reality of entrenched power and the enduring repressive capacity of the powerful. Where others have been enthusiastic, Owen and Morozov suggest caution: the rules of the political game still matter. They are certainly right, and this volume represents my attempt to address the emergence and adoption of new technology in light of what have certainly been a sobering few years.

And what of the relationship between social movements and new technology? In *Digitally Enabled Social Change,* Jennifer Earl and Katrina Kimport document changes to the way advocacy groups organize online, suggesting that new digital repertoires of contention, especially online, push scholars of collective action to focus less on social movements writ large and more on individual acts of protest, regardless of where they occur (i.e., on the streets or online).[2] This observation has lessons for students of civil society more broadly, as it highlights the importance of pivoting from organizational forms to instead focus on collective action and sites of action. Taking this logic one step further suggests we must keep an eye on technology itself if we are to stay with the action, or *stay with the trouble,* as Donna Haraway recently put it.[3]

In the *Logic of Connective Action,* W. Lance Bennett and Alexandra Segerberg split the difference between digital technology's fiercest critics and most ardent fans. Rather than suggesting that new digital tools, like the Internet, have changed everything for contentious politics, or suggesting that the pressures of *realpolitik* ultimately override all other considerations, the authors suggest that a certain amount of hybridity is at play. Their work suggests new digital tools connect people in a way that aggregates power and produces political discourse independent of established organizational resources and irrespective of Mancur Olson's famous free rider problem, which stipulates that people are unlikely to sacrifice much for gains they would enjoy if they did nothing.[4]

Selective incentives would be needed to induce most people to get involved in collective action. It is this logic that Bennett and Segerberg's work turns on its head, as noted on the jacket of their book: "Communication operates as an organizational process that may replace or supplement familiar forms of collective action based on organizational resource mobilization, leadership, and collective-action frames."[5]

Students of social transformation efforts should take note, as Bennett and Segerberg's work suggests: social media have unique emergent properties, rather than simply being faster, cheaper, or broader versions of something we already have (networks, communication channels, media environments, social spaces, and so forth). This point is illustrated by Bennett and Segerberg's development of three different logics of collective action and connective action. Some efforts are organization-brokered, in which a resourced institution takes the lead in mobilizing constituents and the public. Other efforts are organizationally enabled, as when a resourced institution develops a hashtag or offers supporters an online collaboration space. These institutionally brokered and enabled spaces are contrasted with crowd-organized and technologically enabled action. Examples include Alicia Garza's creation of the #BlackLivesMatter hashtag, which catalyzed the eponymous movement after the acquittal of George Zimmerman, and Tarana Burke's creation of #MeToo, which was dormant for a decade before catching fire after the 2017 revelations of abuse by Harvey Weinstein.[6]

These crowd-enabled efforts scale up quickly, mobilize large numbers, dynamically track and target incumbents, and have adaptive repertoires. In this way, social media creates its own logics of engagement, apparently defying the laws of associational gravity articulated by Olson: lower transaction costs mean more people are willing to join up, fewer people engage in free-riding, and fewer people care about those who don't engage.[7] Bennett and Segerberg's work has an important impact on my own field of social movements. As sociologist Brayden King has argued about social media, "sociologists who study social movements have been slow to address their role in activism."[8]

This may be due to a conviction—implicit, perhaps, among sociologists at least—that new digital technologies simply amplify or echo older, well-understood modes of communication. People speak out on Twitter and their voice is amplified on Fox News, but the general effect is the same: lots of people hear messages and then decide what to do about them. Perhaps social movement scholars consider social media to be an ever-accelerating quantitative variable rather than a dramatic and singular qualitative transformation.

An informal review of the major publication venues in social movement scholarship—including journals (*Mobilization* and *Social Movement Studies*) and topical series (Social Movements, Protest, and Contention [now defunct] at the University of Minnesota Press and Contentious Politics at Cambridge University Press)—suggests this may be the case. *Social Movement Studies* has published more articles on social media than has *Mobilization*, but both have specifically focused on technology as a means of mobilization. Books on the topic are thin at Minnesota and Cambridge. The former published Roscigno and Danaher's work on the importance of the radio,[9] as well as Schurman and Munro on activism against biotechnological innovation.[10] Cambridge published Bennett and Stegerberg, but no similar volumes appear in their catalog. The Oxford University Press series on Digital Politics is exemplary, but powered mostly by studies of political communication.

A review of communication scholarship is more revealing, as it becomes clear leading movement scholars[11] have been crossing over to publish important work in journals like *Information, Communication and Society* as well as *New Media & Society*. In those publications, movement scholars join a vibrant community of communication scholars exploring the intersection of new media technologies and collective action. The same can be said of *Social Science Computer Review* and *First Monday*, but not of leading journals in sociology. It may be that social movement scholarship is in a specific kind of denial or doldrums in relation to technology, but I suspect this is an issue within scholarship on civil society, advocacy, and human rights more broadly. This much is suggested by Andrew Chadwick, who argues that mainstream scholarship on political communication has tended to ignore digital media and the Internet, and that the favor has been returned by scholars of the Internet and politics, who have neglected non-Internet media forms and that are unhelpfully "dominated by assumptions about 'revolutionary' change or by a too narrowly drawn frame of 'politics as usual.'"[12]

Scholarship at the intersection of new media and political change has made significant headway in the past decade, the aforementioned challenges notwithstanding,[13] and Bennett and Segerberg make significant contributions to our understanding of critical social processes. Indeed, they are explicit in stating that it is social processes rather than *technology qua technology* that has their attention: "the question here is not whether a particular medium is being used, but how and in what context, by whom, and with what sort of control and conflict within organizations and broader user communities."[14] Here the social, rather than the technical or techno-social, retains pride of place in the causal explanation. Likewise, more work must be done to unpack the relationship between the online and the offline.

This interplay of online and offline is taken up by Andrew Chadwick in *The Hybrid Media System*.[15] Chadwick suggests a "holistic approach to the role of information and communication and politics" is necessary to move scholarly work beyond the false dichotomies of old and new, digital and analog, online and offline.[16] To develop such a holistic approach, Chadwick turns to the concept of hybridity. Hybrid media systems emerge when established broadcast media exist in the same cultural space as snippets shared on social media. The story is not new versus old, but new *and* old evolving simultaneously. But new and old what?

Chadwick's approach to media is of particular utility to the argument I am developing here. The notion of *media logics* is used to accommodate a plurality of forms (i.e., hybridity). By media logics, Chadwick points to "technologies, genres, norms, behaviors, and organizational forms...in the reflexively connected fields of media and politics."[17] This broad-ranged approach considers more, far more perhaps, than particular media tools or spaces and opens his inquiry to a much broader view of what phenomena might be under consideration. The result of this approach can be seen when Chadwick argues that "older media practices in the interpenetrated fields of media and politics adapt and integrate the logics of newer media practices."[18] Central to this approach is a skepticism over the term *new technology*, since, as Carolyn Marvin points out in her book *When Old Technologies Were New*, "*new technologies* is a historically relative term."[19]

Chadwick's work helpfully troubles the space between "new" and "old," confounds efforts to dichotomize the digital and analog, and adds dimensionality to studies that privilege social factors over technology (what Bennett and Segerberg refer to as *medium*). Each of these dichotomies are useful, but for the purposes of this volume, I have chosen to turn them into a continuum and to then put them into conversation with one another.

That is exactly what Steven Livingston and Gregor Walter-Drop do in their recent edited volume, *Bits and Atoms*.[20] Building off of Max Weber's notion that statehood involves a monopoly on violence, the ability to make and enforce rules, and provide public goods, the authors suggest that *limited statehood* describes the absence of these capacities.

Bits and Atoms demonstrates, in case after case, that the digital and the non-digital are running alongside one another in areas of limited statehood and that this is exactly what happens everywhere. While they set themselves a particularly ambitious task of exploring the extent to which new technologies can reliably stand in for the state—as a form of governance—multiple case studies suggest the importance of this question across polities and national contexts. In cases drawn from the Global South—but that also apply worldwide, I would argue—digital tools approximate and patch in for important systems and processes, including public goods that the state is meant to provide. However, sustained and sustainable social, political, and economic life emerges from the complementary interplay of both digital and analog technologies. Here we find both digital communication tools and the older technologies of bureaucracy and infrastructure, to name a particularly important combination, operating in unplanned but patterned ways.

Each of these studies complicates our received understanding of *new digital technology*, pointing instead to the ways that the new intersects with the old and the digital overlaps with the analog. Yet virtually all of these focus on the role of technology in political communication, the clear exception being the work of Livingston and Walter-Drop. For scholars of political and social change, this approach resonates, as it emphasizes the importance of the digital tools increasingly used for communication and mobilization.

However, the task at hand—exploring the politics of tools like drones, satellites, kites, and balloons—requires thinking beyond communication.[21] Scholarship on science and technology is needed if we are to capture not only the way advocacy groups publicize issues, but also to understand the material artifacts people use to realize change. Collective-action efforts rely on, respond to, and operate within important technological and material realities that scholarship on social movements and political communication simply does not cover.

2 AN ALTERNATIVE HISTORY
OF SOCIAL MOVEMENT THEORY

Scholarship on political communication draws—implicitly and explicitly—on the work of luminaries like Marshall McLuhan and Manuel Castells. But what if we instead started our inquiry with Karl Marx? To foreground my argument, I read Marx as a more nuanced observer of the interplay between structure and agency than he is sometimes given credit. Better specifying Marx's recognition of the interplay between these factors highlights the places where agency-, culture-, and emotion-centered approaches to movements may have overcorrected, effectively championing individualistic voluntarism over (rather than *alongside*) structural factors and forces. Keeping a better eye on the dialectical nature of structure and agency—a process which necessarily recognizes the material—may help movement scholars to avoid technological determinism while also focusing attention on technology's social role. The implications are not insignificant. We can ask ourselves what classic social movement theory would look like should it attend to technology, rather than to technology's fruits.[1]

The first intersection between technology and movements lies at the macro level, as broad changes in science and technology shape sociopolitical relationships and opportunities for contentious politics. This line of scholarship traces back to Marx, whose focus on changes in the means of production led to his theory of radical social transformation and political change. This approach was adapted by Charles Tilly in his influential 1978 book *From Mobilization to Revolution.* There Tilly drew on Marxian principles to illustrate the relationship between key movement factors—organization, repression/facilitation, and opportunity/threat—in explaining mobilization and collective action. From whence do these factors spring in Tilly's argument? Not from Marx's organization of production, but instead from a combination of power and interest. The important decision to replace the organization of production (figure 8.1) with interests (figure 8.2) underemphasizes the very real material factors that shape the contexts movements

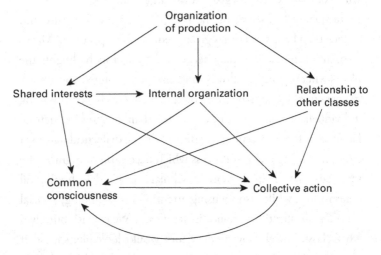

Figure 8.1
Tilly's simple Marxist model ("Organization of Production"). Source: Charles Tilly, *From Mobilization to Revolution* (New York: McGraw-Hill, 1978), 43.

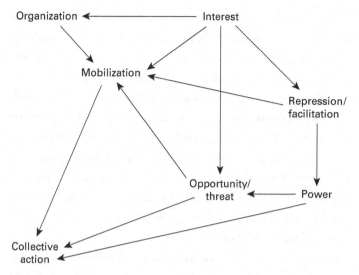

Figure 8.2
Tilly's simple political process model ("Interests"). Source: Charles Tilly, *From Mobilization to Revolution* (New York: McGraw-Hill, 1978), 56.

operate within. This has the effect of reducing consciousness to opportunities and threats (as Tilly would later frame it) and suggesting all paths lead to collective action. Absent is a sense of how interests are turned into action. Also missing is Marx's recognition that consciousness and action have a recursive relationship.[2]

When Doug McAdam set out to refine this approach, he reintroduced the causal importance of science and technology (in the collapse of King Cotton) while incorporating the Weberian emphasis on attribution and appropriation in a secondary role. Echoing Marx in the *18th Brumaire*, incumbents and challengers find themselves operating in conditions beyond their control (broad destabilizing changes), but exercise agency in deciding what to do next (attribution, appropriation). In so doing, McAdam effectively split the difference between Marx and Weber to produce a model extending from the broad destabilizing changes that are themselves rooted in social and economic transitions beyond the movement (figure 8.3).

This approach laid the groundwork for a generation of scholars focused on tracing the moment when communities realize that certain conditions contradict their interests. Surprisingly little work has been done to explain the relationship between these broad changes and those moments of realization. "Political opportunity" thinking emerged in an effort to describe the nature and operation of the broad destabilizing changes that "opened" or "closed" opportunities for challengers. Vibrant debate over what constitutes an open or closed system has evolved into a debate over perception of opportunity or threat, effectively (and deliberately) shifting the arena in question from structure (broad changes in a system) to agency (attribution and appropriation). I am in favor

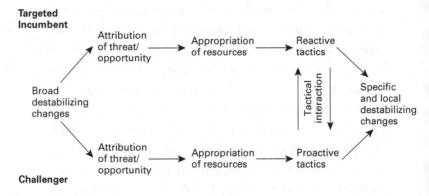

Figure 8.3

McAdam's theory of the onset of contention. Source: Doug McAdam, *Political Process and the Development of Black Insurgency, 1930–1970* (Chicago: University of Chicago Press, 2010).

of these efforts to reclaim an important role for culture and emotion, but believe that a counterbalance is in order.

I am not alone. Andrew Walder has convincingly argued that a focus on the process of mobilization—the marshaling of resources, recruiting of adherents, and navigating of politics—leaves to the side a more pressing question: where do their ideologies, aims, motivations, and tactical choices come from?[3] While this may invite speculation about cultural norms, Walder's answer is that they lie in broader factors and forces, and he advocates for the revival of explanations that draw on the causal power of social structures, as seen in the work of Michael Schwartz and Rory McVeigh.[4]

New work by Jeff Goodwin and Gabriel Hetland suggests that even identity movements and "post-materialist" movements are powerfully shaped by capitalism. Like Walder, they are at pains to remind movement scholars of the importance of scholars who identified the impact that "capitalist dynamics" had on movements (Tilly chief among them). After all, Goodwin and Hetland point out, many classic movement cases revolve around labor mobilization in sites of capitalist expansion and institutions—assembly lines, for example, are sites of both exploitation and collective identity.[5]

A parallel line of scholarship is rooted in the 1974 publication of Immanuel Wallerstein's *The Modern World-System*. Wallerstein suggested changes in the means of production are important, as the world economy is predicated on a division of labor between (1) a hegemonic economic core, (2) a semiperiphery that facilitates transactions, finishes goods, and buffers shocks, and (3) a peripheral zone punctuated by extraction and exploitation.[6]

The means and system of production are *the* primary causal explanation for the situations that marginalized communities find themselves within. Scholars like Michael Schwartz suggest that protest groups develop assessments of the problem and decide on possible solutions based at least in part on their class position, itself tied back to the way in which labor markets are organized within a broader economic context punctuated by forces of production.[7]

My intent is not to resurrect a deterministic approach to technology, but to instead emphasize one way technology might have been more directly incorporated into social movement theories. An alternative history of mainstream movement thinking can perhaps be envisioned if we consider for a moment the way Tilly chose to address the role of technology in contention. In his influential *Regimes and Repertoires,* for example, technology is introduced just long enough to be dismissed on the grounds that technical innovations are subordinate to local political processes, and that "purposes override techniques."[8]

Having disposed of the material by suggesting technological determinism is the only way to conceptualize technologies, and by dismissing technologies as mere "techniques," Tilly introduces political opportunity structures, effectively directing the reader's attention back to the realm of politics. Technologies are a subset of resources, to be sure, but their crucial role in shaping the organization of production is minimized.

Yet Tilly's own argument for why particular repertoires emerged in England between 1750 and 1830 points to a number of factors that are fundamentally rooted in changes in technological innovation and transformations of the means of production, including their concentration, which led to the unique growth of British capitalism[9] and the subsequent concentration of that capital, which then led to the proletariatization of the British workforce.[10] My goal here is not to set aside the key causal roles of political, social, or individual actors, but to instead highlight the enabling and constraining roles technologies play in already-familiar stories. Tilly's own causal argument relies on an understanding of technologies as more than broad enabling environmental factors.

Technology also stalks current theoretical accounts of movement emergence. Early resource-mobilization theorists John McCarthy and Mayer Zald identified the importance of key economic resources to movements.[11] This is usually seen in terms of financial capital, but may also be leadership capacity or technology transfers from elite supporters. These resources have the effect of providing selective incentives for engagement from prospective supporters[12] and also provide crucial institutional infrastructure necessary to pursue key movement goals.[13] What can these resources buy you? Buildings, tables, chairs, computers, posters, websites, busses and bikes, weapons, phones, and faxes, for example. New digital tools and techniques have served as key resources in many contemporary movements, and have considerably reduced the opportunity costs required for engagement.

The literature on issue framing is focused on the process of matching movement claims with social values, such that the movement's issue is actionable and legible to others, including bystanders, targets, and both current and prospective supporters.[14] Here technology makes an appearance, for example, through the media of pamphlets, newspapers, posters, web-pages, Facebook groups, and hashtags. The role of the amplifying organization or institution, so prominent in scholarship in the 1990s, is diminished significantly, since activists can use new digital technologies to more democratically crop, capture, and tag their own images.[15] One form of technology may replace another, as online connective technologies may slowly supplant the importance of bureaucracy and organization (though I find this to be quite unlikely).

The importance of technology in terms of resources and framing goes without saying, and indeed much of this volume has been dedicated to the technologies that make possible the capture, storage, distribution, interpretation, and spread of data. Clearly, tools and technologies are implicit, if unrecognized, in most conventional explanations of collective action. Our attention, however, might helpfully focus one step back in the process, as innovations in science and technology generate broad socioeconomic changes that create new socioeconomic issues while also opening the window of opportunity to new expressions of identity or rights claims, which themselves rely on tools and technologies, some borrowed directly, others invented or hacked to get the job done.[16]

Thinking about innovation directs our attention to the increased industrial capacity in the American North that led to the Great Migration, attracting African American laborers away from the South and in this way further undermining the cotton industry.[17] Such industrial capacity can be thought of in light of even larger historical economic processes driven by technological innovation and turnover in heavy machinery and other industrial equipment.[18]

Materialist projects face stiff resistance from critics who argue that they overlook what makes humans life what it is, especially emotions, culture, and contingency.[19] I am broadly sympathetic to this criticism, as are some targets of this criticism,[20] but have decided to instead tack into the wind and argue that movement scholarship *has not taken technology seriously enough*. Stakeholders on both sides of the structuralist debate overlook the importance of materiality, the thingness of technologies, focusing instead on its supporting role in economic transformations. A sensible first step might involve better interrogating the relationship between macro-economic waves and contentious politics. This would set the stage for subsequent studies to follow the sociopolitical ramifications of whole technical systems or particular technologies along their trajectory.[21]

One could map, for example, the relationship between contentious politics and the emergence and evolution of such technical systems. This approach could prove useful—necessary perhaps—in helping anticipate the kind of contentious politics automation and artificial intelligence will precipitate. To be clear: material forces are not primordial and deterministic. Weber long ago observed that it was a particular set of cultural values that shaped markets, and Karl Polanyi emphasized that the state and political forces lay the groundwork for capitalist economies. Even Marx, often misread as a determinist, leaves room for humans to "make their own history" as they work to shape the pace and direction of technological innovation and adoption.[22]

Social and economic structures do not determine human events, but it would be foolish to think that humans shape history, individually or collectively, on our own terms. Technology is very much part and parcel of the circumstances that are "already given and transmitted from the past," in Marx's terms.[23] The hard work of teasing out the causal role of these structural forces and human agency has bedeviled a long line of social theorists,[24] so I leave to others the more difficult empirical task of better linking these causal mechanisms to contentious politics.[25]

The point should now be clear: incorporating technology into our theorizing raises important questions for anyone interested in explaining or understanding collective action. I close this book in interesting times. Newly invigorated nationalist parties in settled democracies have clearly demonstrated that the status quo is broken. Radical changes in the means of production have not been matched by radical changes in democracies' ability to hold the powerful to account and distribute gains equitably. Quite the inverse is on display: the strong do what they will and the weak suffer what they must. One point I hope to have made in this theoretical afterword is that an understanding of our times requires attention to technology, both in how it may aid people-power mobilization and also in the ways it makes this mobilization necessary.

NOTES

CHAPTER 0

1. This project was only possible with the collaboration of the Hungarian journalism project *Atlatszo*. Special thanks go to Tamas Bodoky, Áron Halász, and Akos Baranya. Critical support was also provided by Eva Bognar at Central European University's Center for Media, Data and Society.

2. Francesca Polletta and Kelsy Kretschmer, "'Free Spaces' in Collective Action," *Theory and Society* 28, no. 1 (1999): 1–38.

3. Erica Chenoweth and Maria J. Stephan, *Why Civil Resistance Works: The Strategic Logic of Nonviolent Conflict* (New York: Columbia University Press, 2011).

4. David Cortright, Rachel Fairhurst, and Kristen Wall, *Drones and the Future of Armed Conflict: Ethical, Legal, and Strategic Implications* (Chicago: University of Chicago Press, 2015); Lisa Parks and Caren Kaplan, *Life in the Age of Drone Warfare* (Durham, NC: Duke University Press, 2017); Hugh Gusterson, *Drone: Remote Control Warfare* (Cambridge, MA: MIT Press, 2016); Ian G. R. Shaw, *Predator Empire: Drone Warfare and Full Spectrum Dominance* (Minneapolis: University of Minnesota Press, 2016); Grégoire Chamayou and Janet Lloyd, *A Theory of the Drone* (New York: The New Press, 2015); Christopher J. Fuller, *See It Shoot It: The Secret History of the CIA's Lethal Drone Program* (New Haven, CT: Yale University Press, 2017); Medea Benjamin, *Drone Warfare: Killing by Remote Control* (New York: Verso Books, 2013).

5. Doug McAdam, *Political Process and the Development of Black Insurgency, 1930–1970* (Chicago: University of Chicago Press, 2010); Sidney G. Tarrow,

Power in Movement: Social Movements and Contentious Politics (New York: Cambridge University Press, 2011); Suzanne Staggenborg, *Social Movements* (New York: Oxford University Press, 2015).

6. Zizi Papacharissi, *Affective Publics: Sentiment, Technology, and Politics* (Oxford: Oxford University Press, 2015).

7. Jennifer Earl and Katrina Kimport, *Digitally Enabled Social Change: Activism in the Internet Age* (Cambridge, MA: MIT Press, 2011).

8. Jessica Lucia Beyer, *Expect Us: Online Communities and Political Mobilization* (New York: Oxford University Press, 2014).

9. Alison Dahl Crossley, "Facebook Feminism: Social Media, Blogs, and New Technologies of Contemporary Us Feminism," *Mobilization:* 20, no. 2 (2015). See also Sarah Gaby and Neal Caren, "Occupy Online: How Cute Old Men and Malcolm X Recruited 400,000 US Users to OWS on Facebook," *Social Movement Studies* 11, no. 3–4 (2012).

10. Bennett and Segerberg, *The Logic of Connective Action.*

11. Andrew T. Little, "Communication Technology and Protest," *The Journal of Politics* 78, no. 1 (2016); Jan H. Pierskalla and Florian M. Hollenbach, "Technology and Collective Action: The Effect of Cell Phone Coverage on Political Violence in Africa," *American Political Science Review* 107, no. 2 (2013).

12. Beyer, *Expect Us.*

13. Bennett and Segerberg, *The Logic of Connective Action.*

14. Polletta and Kretschmer, "'Free Spaces.'"

15. Evgeny Morozov, *To Save Everything, Click Here: The Folly of Technological Solutionism* (New York: PublicAffairs, 2013).

16. Woolley, Samuel C. and Philip N. Howard, eds., *Computational Propaganda: Political Parties, Politicians, and Political Manipulation on Social Media.* (Oxford, UK: Oxford University Press, 2018).

17. Owen, *Disruptive Power: The Crisis of the State in the Digital Age* (New York: Oxford University Press, 2015).

18. Earl and Kimport, *Digitally Enabled Social Change.*

19. Bennett and Segerberg, *The Logic of Connective Action.*

20. Andrew Chadwick, *The Hybrid Media System: Politics and Power* (New York: Oxford University Press, 2017).

21. Steven Livingston and Gregor Walter-Drop, *Bits and Atoms: Information and Communication Technology in Areas of Limited Statehood* (New York: Oxford University Press, 2013).

22. Zeynep Tufekci, *Twitter and Tear Gas: The Power and Fragility of Networked Protest* (New Haven, CT: Yale University Press, 2017).

23. Clifford Bob, *The Global Right Wing and the Clash of World Politics* (New York: Cambridge University Press, 2012).

24. Patrick Meier, *Digital Humanitarians: How Big Data Is Changing the Face of Humanitarian Response* (Boca Raton, FL: CRC Press, 2015).

25. Livingston and Walter-Drop, *Bits and Atoms.*

26. Sheila Jasanoff, *States of Knowledge: The Co-Production of Science and the Social Order* (New York: Routledge, 2004); Paul Dourish, *The Stuff of Bits: An Essay on the Materialities of Information* (Cambridge, MA: MIT Press, 2017); Lisa Parks and Nicole Starosielski, *Signal Traffic: Critical Studies of Media Infrastructures* (University of Illinois Press, 2015); Nicole Starosielski, *The Undersea Network* (Durham, NC: Duke University Press, 2015).

27. Indeed, this is the definition used by movement scholar David A. Snow; see "Social Movements as Challenges to Authority: Resistance to an Emerging Conceptual Hegemony," in *Authority in Contention,* ed. Daniel Cress and Daniel J Myers (Emerald Group Publishing, 2004), 11.

28. Owen, *Disruptive Power,* 14–18.

29. David J. Hess, *Undone Science: Social Movements, Mobilized Publics, and Industrial Transitions* (Cambridge, MA: MIT Press, 2016); Scott Frickel et al., "Undone Science: Charting Social Movement and Civil Society Challenges to Research Agenda Setting," *Science, Technology, & Human Values* 35, no. 4 (2010): 444–467.

30. David Edgerton, *The Shock of the Old: Technology and Global History since 1900* (New York: Oxford University Press, 2007). For those keeping track, I follow Vygotsky (1980) in broadly defining tools as having both psychological as well as material components. While the former allows for complex cognitive functions (Cole and Wertsch 1996), the latter describes my usage throughout this volume: paper, rock, scissors, but also apps and algorithms.

My adoption of this approach is rooted in an appreciation for the extent to which Vygotsky positions tools as making possible human creativity and action. Kallinikos, Leonardi, and Nardi (2012, 10) push this argument further, suggesting structure allows action "rather than being simply constrained by structure, as the typical conventional interpretive understanding wants us to believe, human choice and agency are made originally possible through the very resources that objects and structures dispose." The structure-agency dichotomy is a false one, then, since structures themselves are sites of action. If I were a more playful sort, I would follow Donna J. Haraway's approach, and dub all I find *kin* of some sort or another.

31. Noortje Marres and Javier Lezaun, "Materials and Devices of the Public: An Introduction," *Economy and Society* 40, no. 4 (2011): 3, 6.

32. Langdon Winner, *The Whale and the Reactor: A Search for Limits in an Age of High Technology* (Chicago: University of Chicago Press, 2010), 22.

33. If I were a stickler for terminology, which I guess I am because I have created this endnote, I would have done some work to better incorporate Winner's original definition, as found in his 1978 book *Autonomous Technology: Technics-out-of-Control as a Theme in Political Thought* (Cambridge, MA: MIT Press, 1978, 8–12). There he suggested technology could be thought of as being variously comprised of:

    apparatus: objects (tools, instruments, gadgets)
    technique: technical activities (skill, procedures)
    organization: technical social arrangements (factory, team, etc.)
    network: large-scale systems.

    This approach appeals to me for its inclusion of technique, an approach that I find useful in conceptualizing digital and virtual tools like code and speech, both of which are technologies in their own right. My adoption of the term *stuff* might be too casual for some readers, but I have chosen it in an attempt to signal this broader range of instantiations.

34. Tarleton Gillespie, Pablo J. Boczkowski, and Kirsten A. Foot, *Media Technologies: Essays on Communication, Materiality, and Society* (Cambridge, MA: MIT Press, 2014), 1; emphasis mine.

35. Brandon Vaidyanathan et al., "Causality in Contemporary American Sociology: An Empirical Assessment and Critique," *Journal for the Theory of Social Behaviour* 46, no. 1 (2016): 3–26.

CHAPTER 1

1. Jasanoff, *States of Knowledge.*

2. Thomas Parke Hughes, *Human-Built World: How to Think About Technology and Culture* (Chicago: University of Chicago Press, 2004).

3. Jane Bennett et al., *New Materialisms: Ontology, Agency, and Politics* (Durham, NC: Duke University Press, 2010).

4. Jannis Kallinikos, Paul M. Leonardi, and Bonnie A. Nardi, "The Challenge of Materiality: Origins, Scope, and Prospects," in *Materiality and Organizing: Social Interaction in a Technological World,* ed. Paul M. Leonardi, Bonnie A. Nardi, and Jannis Kallinikos (Oxford, UK: Oxford University Press, 2012), 3–22.

5. Livingston and Walter-Drop, *Bits and Atoms.*

6. Parks and Starosielski, *Signal Traffic.*

7. Starosielski, *The Undersea Network.*

8. Dourish, *The Stuff of Bits.*

9. Pablo Boczkowski and Leah A. Lievrouw, "Bridging STS and Communication Studies: Scholarship on Media and Information Technologies," *The Handbook of Science and Technology Studies,* 3rd ed., ed. Edward J. Hackett, Olga Amsterdamska, Michael Lynch, and Judy Wajcman (Cambridge, MA: MIT Press, 2008), 967.

10. Lourdes A Vera et al., "Data Resistance: A Social Movement Organizational Autoethnography of the Environmental Data and Governance Initiative," *Mobilization,* 23, no. 4 (2018): 511–529; Toly Rinberg et al., "Changing the Digital Climate: How Climate Change Web Content is Being Censored Under the Trump Administration" (Environmental Data and Governance Initiative, 2018); Toly Rinberg and Andrew Bergman, "Censoring Climate Change," *The New York Times,* November 22, 2017, https://www.nytimes.com/2017/11/22/opinion/censoring-climate-change.html.

11. Frances Fox Piven and Richard A. Cloward, *Poor People's Movements: Why They Succeed, How They Fail* (New York: Vintage Books, 1979).

12. Chenoweth and Stephan, *Why Civil Resistance Works.*

13. James C. Scott, *Weapons of the Weak: Everyday Forms of Peasant Resistance* (New Haven, CT: Yale University Press, 2008).

14. I would like to thank Dexter Pratt for pointing out that the prose gets decidedly technical here. I dedicate this detour paragraph to him.

15. Charles Tilly, *Regimes and Repertoires* (Chicago: University of Chicago Press, 2010).

16. Neil Fligstein and Doug McAdam, "Toward a General Theory of Strategic Action Fields," *Sociological Theory* 29, no. 1 (2011): 1–26.

17. Terence E. McDonnell, *Best Laid Plans: Cultural Entropy and the Unraveling of Aids Media Campaigns* (Chicago: University of Chicago Press, 2016).

18. Thanks to Lars Almquist for the mosquito net examples from his fieldwork in Burundi.

19. Nick Riggle, *On Being Awesome: A Unified Theory of How Not to Suck* (New York: Penguin, 2017).

20. Austin Choi-Fitzpatrick, *What Slaveholders Think: How Contemporary Perpetrators Rationalize What They Do* (New York: Columbia University Press, 2017).

21. Thomas R. Rochon, *Culture Moves: Ideas, Activism, and Changing Values* (Princeton, NJ: Princeton University Press, 2000), 1.

22. Gene Sharp, *The Politics of Nonviolent Action,* 3 vols. (Boston: Porter Sargent, 1973).

23. "198 Methods of Nonviolent Interaction," The Albert Einstein Institution, Boston, http://www.aeinstein.org/wp-content/uploads/2013/09/198_methods -1.pdf.

24. Paul Burstein, "The Impact of Public Opinion on Public Policy: A Review and an Agenda," *Political Research Quarterly* 56, no. 1 (2003): 29–40; John D. McCarthy and Mayer N. Zald, "Resource Mobilization and Social Movements: A Partial Theory," *American Journal of Sociology* (1977): 1212–1241.

25. Charles Tilly, "Spaces of Contention," *Mobilization,* 5, no. 2 (2000): 35.

26. Dennis Chong, "Pavement Glued Down in Hong Kong for China Official Visit," AFP, May 16, 2016.

27. Tilly, *Regimes and Repertoires,* 56–57.

28. Benedict Anderson, *Imagined Communities: Reflections on the Origin and Spread of Nationalism* (New York: Verso Books, 2006).

29. James J. Gibson, *The Ecological Approach to Visual Perception: Classic Edition* (Psychology Press, 2014).

30. Ian Hutchby, "Technologies, Texts and Affordances," *Sociology* 35, no. 2 (2001): 441–456.

31. Peter Nagy and Gina Neff, "Imagined Affordance: Reconstructing a Keyword for Communication Theory," *Social Media + Society* 1, no. 2 (2015); McDonnell, *Best Laid Plans*; Terence E. McDonnell, "Cultural Objects as Objects: Materiality, Urban Space, and the Interpretation of Aids Campaigns in Accra, Ghana," *American Journal of Sociology 115*, no. 6 (2010): 1800–1852.

32. Wiebe E. Bijker et al., *The Social Construction of Technological Systems: New Directions in the Sociology and History of Technology* (Cambridge, MA: MIT Press, 2012).

33. Hutchby, "Technologies, Texts and Affordances."

34. Ibid.

35. Bruno Latour, *Reassembling the Social: An Introduction to Actor-Network-Theory* (New York: Oxford University Press, 2005), 316.

36. Wiebe Bijker and John Law, *Shaping Technology/Building Society: Studies in Sociotechnical Change* (Cambridge, MA: MIT Press, 1994), 8.

37. Earl and Kimport, *Digitally Enabled Social Change*, 132.

38. Bennett and Segerberg, *The Logic of Connective Action*.

39. Steven Livingston, "Digital Affordances and Human Rights Advocacy" (Berlin: DFG Collaborative Research Center, 2016), 10.

40. Tilly, *Regimes and Repertoires*, 43, fn. 3.

41. Ian Hutchby, *Conversation and Technology: From the Telephone to the Internet* (Hoboken, NJ: John Wiley & Sons, 2013).

42. William W. Gaver, "Technology Affordances," in *Proceedings of the SIGCHI Conference on Human Factors in Computing Systems* (New York: ACM, 1991), 139–146.

43. Hutchby, "Technologies, Texts and Affordances," 443.

44. This observation is true at levels additionally identified by Georg Simmel and Martin Buber.

45. Gina Neff et al., "Affordances, Technical Agency, and the Politics of Technologies of Cultural Production" (Broadcast Education Association, 2012): 299–313.

46. Ibid., 26.

47. Jenny L. Davis and James B. Chouinard, "Theorizing Affordances: From Request to Refuse," *Bulletin of Science, Technology & Society* 36, no. 4 (2016): 1.

48. Victor Kaptelinin and Bonnie A. Nardi, *Acting with Technology: Activity Theory and Interaction Design* (Cambridge, MA: MIT Press, 2006), 32–33.

49. I'm indebted to Andrew R. Brown's "Understanding Musical Practices as Agency Networks" (*Proceedings of the Seventh International Conference on Computational Creativity*, Paris: Association of Computational Creativity, 2016) for this particular formulation of their thinking.

50. Students of social theory should take note, as this approach directly contradicts efforts by scholars like Mustafa Emirbayer and Ann Mische, who suggest that people are the only actors with causal significance in the social world; see Mustafa Emirbayer and Ann Mische, "What Is Agency?" *American Journal of Sociology* 103, no. 4 (1998): 962–1023; James V. Wertsch, *Mind as Action* (New York: Oxford University Press, 1998).

51. Kaptelinin and Nardi, *Acting with Technology*, 242.

52. For more on emergent properties in complex systems see Yaneer Bar-Yam, *Dynamics of Complex Systems* (New York: Routledge, 2018).

53. Ibid., 280. Kaptelinin and Nardi draw on the work of Craig W. Reynolds, "Flocks, Herds and Schools: A Distributed Behavioral Model" (paper presented at the ACM SIGGRAPH Computer Graphics Conference, New York: ACM, 1987), and readers may also consult Robin Hanson, *The Age of Em: Work, Love, and Life When Robots Rule the Earth* (New York: Oxford University Press, 2016) and Nick Bostrom, *Superintelligence: Paths, Dangers, Strategies* (New York: Oxford University Press, 2014).

54. Bostrom, *Superintelligence*.

55. By emergent phenomena, I simply mean unanticipated recombination of activities that do not conform to original human-originated intentions.

56. Peter Warren Singer, *Wired for War: The Robotics Revolution and Conflict in the 21st Century* (New York: Penguin, 2009).

57. In her book *Disaster Robotics* (Cambridge, MA: MIT Press, 2014, 11), Roboticists Without Borders founder and Texas A&M professor Robin R. Murphy suggests that robots (including UAVs) operate where people and other animals cannot.

58. Winner, *The Whale and the Reactor*, 13.

59. Austin Choi-Fitzpatrick, "Drones for Good: Technological Innovations, Social Movements, and the State," *Journal of International Affairs* 68, no. 1 (2014): 19.

60. I recognize that my definitions diverge from those offered by scholars of, for example, innovative disruption, where the focus is on the creation of new markets and the unlocking of new value networks in such a way that incumbents are displaced. Likewise, the term emergent is less often applied to technologies than to science.

CHAPTER 2

1. This section draws on a 2012 profile of Karem in the *Washington Post*: Peter Finn, "Rise of the Drone: From Calif. Garage to Multibillion-Dollar Defense Industry," December 23, 2011, https://www.washingtonpost.com/national /national-security/rise-of-the-drone-from-calif-garage-to-multibillion-dollar -defense-industry/2011/12/22/gIQACG8UEP_story.html.

2. Benjamin S. Lambeth, *NATO's Air War for Kosovo: A Strategic and Operational Assessment* (Santa Monica, CA: Rand Corporation, 2001).

3. Paul Virilio and Benjamin H. Bratton, *Speed and Politics* (Los Angeles: Semiotext(e), 2006), 90.

4. Indeed, by the late 1990s, the Gnat had a new home in General Atomic and a new role as a hunter rather than an observer. Yet his insistence that it wasn't he who armed the device should not be mistaken for disapproval. In our conversation, he essentially argued that new wars call for new tools and that the surveillance requirements of the Cold War were linked to the threats of large stockpiles. Arming the thing would have turned the Cold War hot, he implied.

5. Many excellent books have been written about the promise and peril of UAVs in battlespace. The reader would be well advised to consult Cortright, Fairhurst, and Wall, *Drones and the Future of Armed Conflict*.

6. Richard Whittle, *Predator: The Secret Origins of the Drone Revolution* (New York: Macmillan, 2014), 83. A team of CIA and General Atomics personnel flew the Gnat 750 over Sarajevo from a location in Albania 140 miles away. By 1994, GPS had been added to the platform, and military drones were developed through the late 1990s. The rest is history.

7. Rehman, Rafiq ur, "Please Tell Me, Mr President, Why a US Drone Assassinated My Mother," Reuters, October 25, 2013, https://www.theguardian.com/commentisfree/2013/oct/25/president-us-assassinated-mother.

8. This collaboration has resulted in a number of publications, including Gordon Hoople, Austin Choi-Fitzpatrick, and Elizabeth Reddy, "Drones for Good: Interdisciplinary Project Based Learning Between Engineering and Peace Studies," *International Journal of Engineering Education* 35, no. 2 (2019): 1378–1391; Elizabeth Reddy, Gordon Hoople, and Austin Choi-Fitzpatrick, "Boundary Objects in Classroom Practice: Using Drones to Foster Critical Engagement," *Journal of Engineering Studies* 11, no. 1 (2019): 51–64.

9. Hutchby, "Technologies, Texts and Affordances," 447.

10. While these latter two technologies are also geospatial affordances, I have done without them for two reasons. The first is that there is already a vast literature on their use and role starting from the moment of their invention. The second reason is that despite decades of fiddling, both platforms remain complex and expensive. There is no reason to anticipate they will be democratized in a manner that makes them accessible to the general public.

11. Livingston, "Digital Affordances and Human Rights Advocacy," 10. Where Livingston sees this as a digital affordance (alongside digitally networked affordances and forensic affordances), I would like to remain agnostic about whether a given technology is digital or analog, focusing instead on the opportunities they offer for action in the air.

12. Livingston, "Digital Affordances and Human Rights Advocacy," 9.

13. Ibid.

14. Robert D. Benford and David A. Snow, "Framing Processes and Social Movements: An Overview and Assessment," *Annual Review of Sociology* (2000): 611–639.

15. Frickel et al., "Undone Science: Charting Social Movement and Civil Society Challenges to Research Agenda Setting"; Hess, *Undone Science*.

16. Pierre Bélanger, "Altitudes of Urbanization," *Tunnelling and Underground Space Technology Incorporating Trenchless Technology Research*, no. 55 (2016).

17. Nora K. Chadwick, "The Kite: A Study in Polynesian Tradition," *The Journal of the Royal Anthropological Institute of Great Britain and Ireland* 61 (1931): 455–491.

18. There are fewer scholars of the kite than there ought to be. This chapter relies heavily on the work of Clive Hart, whose *Kites, An Historical Survey* (Mount Vernon, NY: Paul P. Appel, 1967; repr., 1982) remains authoritative. I have also found Lee Newman and Jay Hartley Newman, *Kite Craft: The History and Processes of Kitemaking Throughout the World* (New York: Crown, 1988) quite useful.

19. Joseph Needham, *Science and Civilisation in China; Vol 4. Physics and Physical Technology. Part II Mechanical Engineering* (London: Cambridge University Press, 1965), 577, fn. A.

20. Choe Sang-su, "The Survey of Korean Kites" (Seoul: Korea Books Pub. Co, 1958) is quoted in Hart, *Kites, an Historical Survey*, 34.

21. Hart's *Kites, an Historical Survey* (see especially 61–63) doubts this is the case, though I suspect it is because he appears determined to sketch an alternative origin story for the European kite. Rather than linking it to the Chinese kite, he instead traces its origins to the banners and windsocks of heraldry (something the Chinese do not appear to have had), thereby establishing, wittingly or no, a European claim to the invention. Bollocks, I say.

22. K. Gandhi, "Charles Ellet, Jr., the Pioneer American Suspension Bridge Builder" (paper presented at the *Sustainable Bridge Structures*: Proceedings of the 8th New York City Bridge Conference, New York City, 2015).

23. Michael Brian Schiffer, *Draw the Lightning Down: Benjamin Franklin and Electrical Technology in the Age of Enlightenment* (Berkeley: University of California Press, 2006). While there is some debate over the veracity of Franklin's story, the fact that kites were used to demonstrate the electrical property of lightning is beyond dispute. A number of scholars point to the fact that de Romas demonstrated the same feat on live animals, leaving the audience with little doubt of the demonstration's veracity. See, for example, Hart, *Kites, an Historical Survey*, 94–103, and Juan Miguel Suay and David Teira, "Kites," *Nuncius* 29, no. 2 (2014): 439–463.

24. Hart, *Kites, an Historical Survey*.

25. The flight was originally planned with a balloon, but it was lost in a gale. See William John Baker, *A History of the Marconi Company 1874–1965* (New York: Routledge, 2013).

26. Ibid.

27. Suay and Teira, "Kites."

28. Ibid., 452.

29. This text originally read "desire for flight is a disease with many symptoms." Josie Siegel has inspired me to overhaul my metaphors more broadly, and in this matter suggested that perhaps "all pathology is an opportunity."

30. For more on "man-lifting" kites, see Newman and Newman, *Kite Craft*.

31. George Pocock, *The Aeropleustic Art or Navigation in the Air, by the Use of Kites or Buoyant Sails* (UK: Sherwood & Company, 1827), 19–22; quoted in Hart, *Kites, an Historical Survey*, 115.

32. And here too we must note that Cody and his contemporaries were hard at work on something the Chinese likely landed on first. See Hart, *Kites, an Historical Survey*, 27.

33. David M. Halbfinger, "Unable to Stop Flaming Kites, Israel Moves to Choke Off Gaza Commerce," *The New York Times*, July 9, 2018, https://www.nytimes .com/2018/07/09/world/middleeast/gaza-israel-kites-embargo.html.

34. David M. Halbfinger, "At Gaza Protests: Kites, Drones, Gas, Guns and the Occasional Bomb," *The New York Times*, June 8, 2018, https://www.nytimes .com/2018/06/08/world/middleeast/gaza-israel-protest-fence-border.html.

35. Ling, *Science and Civilisation in China*, 577–578.

36. This anecdote comes from Hart, *Kites, an Historical Survey*, 172–173, who draws from an 1864 article in *Scientific American*.

37. I thank Claire Bergstresser for pointing me in this direction.

38. "Britain Puts Ban on Kite Flying as Safety Move," *New York Herald Tribune*, July 6, 1940, https://iht-retrospective.blogs.nytimes.com/2015/07/05/1940 -britain-puts-ban-on-kite-flying-as-safety-move.

39. Like many innovations, it appears China may have gotten here first. Like gunpowder, China lays claim to the invention itself, but the first recorded

development of the technology at scale occurred elsewhere. This section draws on material broadly available to the public (Wikipedia, hobbyist websites, and so forth), as well as on more scholarly efforts, including Tom D. Crouch, *Lighter Than Air: An Illustrated History of Balloons and Airships* (Baltimore, MD: Johns Hopkins University Press, 2009); Charles Coulston Gillispie, *The Montgolfier Brothers and the Invention of Aviation 1783–1784: With a Word on the Importance of Ballooning for the Science of Heat and the Art of Building Railroads* (Princeton, NJ: Princeton University Press, 2014); Christopher Sprigg, *The Airship—Its Design, History, Operation and Future* (London: Sampson Low, Marsdon & Company, 1931); Ernst August Lehmann et al., *Zeppelin: The Story of Lighter-Than-Air Craft* (London: Longmans, Green and Company, 1937); and Michael John Haddrick Taylor, *Brassey's World Aircraft & Systems Directory, 1999/2000* (London: Brassey's, 1999).

40. The Brazilian-born priest Bartolomeu de Gusmão deserves credit for pioneering work in lighter-than-air flight, as he was reported to have set aloft a paper balloon in the court of Portugal's King John V in 1709, a feat which earned him an academic position and possibly later attracted the ire of the Inquisition.

41. Paul Virilio and Philippe Petit, *Politics of the Very Worst*, trans. Sylvere Lotringer (New York: Semiotext(e), 1999), 89; also see Paul Virilio, *The Vision Machine* (Bloomington: Indiana University Press, 1994).

42. Sprigg, *The Airship*. Lehmann et al., *Zeppelin*, 326.

43. Taylor, *Brassey's World Aircraft & Systems Directory*, 157.

44. Jeffrey Yoo Warren, "Grassroots Mapping: Tools for Participatory and Activist Cartography" (thesis, Massachusetts Institute of Technology, 2010), 71.

45. Whitacre, Andrew. 2010. "Balloon Mapping the Oil Spill Proves Responsive, Open Source," MediaShift, May 19, 2010, http://mediashift.org/2010/05 /balloon-mapping-the-oil-spill-proves-responsive-open-source138.

46. Meier, *Digital Humanitarians*, 83.

47. Warren, "Grassroots Mapping," 70.

48. Shannon Dosemagen, Jeffrey Warren, and Sara Wylie, "Grassroots Mapping: Creating a Participatory Map-Making Process Centered on Discourse," *Journal of Aesthetics and Protest* 8 (2011).

49. I first came across Public Lab while on a panel about new ways of seeing human rights violations. Claudia Martinez Mansell presented very interesting work on

the balloon-based community mapping she was doing with balloons. Together with residents in at-risk and low-resource spaces, she would use a camera-equipped balloon to generate high-quality maps. See Claudia Martinez Mansell, "A Change of Perspective: Aerial Photography and 'the Right to the City' in a Palestinian Refugee Camp," in *Visual Imagery and Human Rights Practice* (New York: Palgrave Macmillan, 2018), 213–228.

50. Charles M. Evans, *The War of the Aeronauts: A History of Ballooning During the Civil War* (Mechanicsburg, PA: Stackpole Books, 2002).

51. Christopher Cole and E. F. Cheesman, *The Air Defence of Great Britain 1914–1918* (London: Putnam, 1984), 449.

52. Douglas Hill Robinson, *Giants in the Sky: A History of the Rigid Airship* (Seattle: University of Washington Press, 1973), 126–127; A. M. Ventry and E. M. Kolesnik, *Airship Saga: The History of Airships Seen through the Eyes of the Men Who Designed, Built, and Flew Them* (Poole, Dorset, UK: Blandford Press, 1982).

53. This is not to discount the fact that military researchers continue to be involved in the kind of meteorological studies that rely on high-altitude balloons up through the present.

54. This section draws on interviews with my colleagues at the University of Nottingham, especially Stuart Marsh and Doreen Boyd.

55. Fredric Jameson and Masao Miyoshi, *The Cultures of Globalization* (Durham, NC: Duke University Press, 1998), xvi.

56. Lisa Parks, *Cultures in Orbit: Satellites and the Televisual* (Durham, NC: Duke University Press, 2005).

57. While it's too early to tell, it may be that the increased presence of individual actors—from Elon Musk to high school students launching CubeSats—heralds the beginning of a fourth phase.

58. Laura Kurgan, *Close Up at a Distance: Mapping, Technology, and Politics* (Cambridge, MA: MIT Press, 2013).

59. National Research Council, *People and Pixels: Linking Remote Sensing and Social Science* (National Academies Press, 1998).

60. Andrew Herscher, "Surveillant Witnessing: Satellite Imagery and the Visual Politics of Human Rights," *Public Culture* 26, no. 3, 74 (2014): 469–500; Lars Bromley, "Eye in the Sky: Monitoring Human Rights Abuses Using Geospatial Technology," *Georgetown Journal of International Affairs* (2009):

159–168; Lisa Parks, "Digging into Google Earth: An Analysis of 'Crisis in Darfur,'" *Geoforum* 40, no. 4 (2009): 535–545.

61. David Hawk, *The Hidden Gulag: Exposing North Korea's Prison Camps* (Washington, DC: US Committee for Human Rights in North Korea, 2003).

62. Committee for Human Rights in North Korea, "HRNK Launches David Hawk's The Parallel Gulag," October 25, 2017, https://www.hrnk.org/events /announcements-view.php?id=45.

63. Anna Fifield, "New Images Show North Korea's Extensive Network of 'Reeducation' Camps," *Washington Post,* October 25, 2017.

64. Stephen Graham, *Vertical: The City from Satellites to Bunkers* (New York: Verso Books, 2016), 25.

65. Ibid., 27.

66. Wolfgang Sachs, in *Planet Dialectics: Explorations in Environment and Development* (London: Zed Books, 2015), on pages 110–111, quotes Peter Sloterdijk, *Versprechen Auf Deutsch*, vol. 631 (Frankfurt: Suhrkamp, 1990), which is available only, to the best of my knowledge, in German.

67. Graham, *Vertical,* 45–51.

68. Ibid.

69. "Quadcopter," Wikipedia, https://en.wikipedia.org/wiki/Quadcopter#cite_ note-dasc04-1.

70. Microdones company story, https://www.microdrones.com/en/team/our -story.

71. "Parrot SA," Wikipedia, https://en.wikipedia.org/wiki/Parrot_SA.

72. DJI about page, https://www.dji.com/company.

73. Austin Choi-Fitzpatrick et al., *Up in the Air: A Global Estimate of Non-Violent Drone Use 2009–2015* (2016), https://doi.org/10.22371/08.2016.001.

74. Cases of sousveillance (the powerless surveilling the powerful) were coded as both "surveillance" and as "advocacy."

75. Cortright, Fairhurst, and Wall, *Drones and the Future of Armed Conflict*; Parks and Kaplan, *Life in the Age of Drone Warfare*; Gusterson, *Drone: Remote Control Warfare*; Shaw, *Predator Empire*; Chamayou and Lloyd, *A Theory of the Drone*; Fuller, *See It Shoot It*; Benjamin, *Drone Warfare*.

76. Pierre Bélanger, "Altitudes of Urbanization."

77. Ibid., 4.

78. Ibid., 4.

79. R. Buckminster Fuller, *Operating Manual for Spaceship Earth* (original publication 1923; estate of R. Buckminster Fuller, 2008).

80. James C. Scott, *Seeing Like a State: How Certain Schemes to Improve the Human Condition Have Failed* (New Haven, CT: Yale University Press, 1998).

81. Eyal Weizman, *Hollow Land: Israel's Architecture of Occupation* (New York: Verso Books, 2012).

82. Michael Edwards, *Civil Society* (Cambridge, UK: Polity, 2009).

83. Nancy Fraser, "Transnationalizing the Public Sphere: On the Legitimacy and Efficacy of Public Opinion in a Post-Westphalian World," *Theory, Culture & Society* 24, no. 4 (2007), 22.

84. Graham, *Vertical*, emphasis added.

85. Henri Lefebvre and Donald Nicholson-Smith, *The Production of Space*, vol. 142 (New York: Oxford Blackwell, 1991), 337.

86. Weizman, *Hollow Land*. See also Rafi Segal, David Tartakover, and Eyal Weizman, *A Civilian Occupation: The Politics of Israeli Architecture* (New York: Verso, 2003) and Stuart Elden, "Secure the Volume: Vertical Geopolitics and the Depth of Power," *Political Geography* 34 (2013): 35–51.

87. Peter Sloterdijk, "Airquakes," *Environment and Planning D: Society and Space* 27, no. 1 (2009): 41–57; *Terror from the Air* (Los Angeles: Semiotext(e), 2009).

88. Paul Virilio, *War and Cinema: The Logistics of Perception* (New York: Verso, 1989); Elden, "Secure the Volume"; Stephen Graham, "The Urban 'Battlespace,'" *Theory, Culture & Society* 26, no. 7–8 (2009): 278–288.

89. Peter Adey, *Aerial Life: Spaces, Mobilities, Affects* (John Wiley & Sons, 2010), 9.

90. Francisco Klauser and Silvana Pedrozo, "Power and Space in the Drone Age: A Literature Review and Politico-Geographical Research Agenda," *Geographica Helvetica* 70, no. 4 (2015): 287.

91. Ibid., 287. Also see Paul Virilio, *The Information Bomb,* trans. Chris Turner (New York: Verso, 2000).

92. Klauser and Pedrozo, "Power and Space in the Drone Age," 287. See also David Lyon, *The Electronic Eye: The Rise of Surveillance Society* (Minneapolis: University of Minnesota Press, 1994); William G. Staples, *Everyday Surveillance: Vigilance and Visibility in Postmodern Life*, 2d ed. (Lanham, MD: Rowman & Littlefield, 2014).

93. Klauser and Pedrozo, "Power and Space in the Drone Age," 287–288.

94. Ibid.

95. Ibid.

96. Graham, *Vertical*, 5–6.

97. Fraser, "Transnationalizing the Public Spheres"; and Nancy Fraser, *Scales of Justice: Reimagining Political Space in a Globalizing World* (New York: Columbia University Press, 2009).

98. Jürgen Habermas, *The Theory of Communicative Action*, 2 vols. (Boston: Beacon Press, 1984), 31–32.

99. Cf. Fraser, *Scales of Justice*, 664.

100. Tufekci, *Twitter and Tear Gas*, 11.

101. Winner, *Autonomous Technology*, 100, emphasis original.

102. Donna Haraway, *Simians, Cyborgs, and Women: The Reinvention of Nature* (New York: Routledge, 2013), 191–192.

CHAPTER 3

1. This chapter draws on case studies of civil society groups' drone use, developed and published with my student coauthors (Choi-Fitzpatrick et al. 2016).

2. Thomas Snitch, "Satellites, Mathematics and Drones Take Down Poachers in Africa," *The Conversation*, January 27, 2015, http://theconversation.com/satellites-mathematics-and-drones-take-down-poachers-in-africa-36638.

3. Ellen Barry, "Rally Defying Putin's Party Draws Tens of Thousands," *The New York Times*, December 10, 2011, http://www.nytimes.com/2011/12/11/world/europe/thousands-protest-in-moscow-russia-in-defiance-of-putin.html.

4. "More on the Moscow Protest Photos," Drone Journalism Lab, December 12, 2011, http://www.dronejournalismlab.org/post/14136093865/more-on-the-moscow-protest-photos.

5. "Aerial Drone Captures Stunning Video Of Massive Ukraine Protest," *Huffington Post*, December 17, 2013, http://www.huffingtonpost.com/2013/12/16/kiev-protest-video-drone-ukraine_n_4455340.html.

6. Ryan Gallagher, "Video Reportedly Shows Drone Shot Down by Police at Turkish Protests," *Slate*, June 24, 2013, http://www.slate.com/blogs/future_tense/2013/06/24/turkish_protests_police_reportedly_shoot_down_demonstrators_drone_video.html.

7. Faine Greenwood, "Thailand Is Cracking Down on Drones," *Slate*, February 5, 2015, http://www.slate.com/articles/technology/future_tense/2015/02/thailand_drone_regulations_why_you_should_care.html.

8. Jonah Fisher, "Thailand Protests: PM Rejects Resignation Demand," *BBC News*, December 2, 2013. http://www.bbc.com/news/world-asia-25181806.

9. Tekendra Parmar, "Drones in Southeast Asia," Center for the Study of the Drone at Bard College, August 14, 2015, http://dronecenter.bard.edu/drones-in-southeast-asia.

10. Chenoweth and Stephan, *Why Civil Resistance Works*; Michael Biggs, "Size Matters: Quantifying Protest by Counting Participants," *Sociological Methods & Research* (2016): 351–383.

11. Choi-Fitzpatrick and Juskauskas, "Up in the Air."

12. Dean Starkman, *The Watchdog That Didn't Bark: The Financial Crisis and the Disappearance of Investigative Reporting* (New York: Columbia University Press, 2014).

13. Nilanjana Bhowmich, "In Unpredictable India, Security Services Embrace the Drone Revolution," *Time*, August 6, 2014, http://time.com/3084645/india-drones-security-humanitarian.

14. Monica Sarkar, "Security from the Sky: Indian City to Use Pepper-Spray Drones for Crowd Control," *CNN*, April 9, 2015, http://www.cnn.com/2015/04/09/asia/india-police-drones.

15. Nadia Khomami, "'Abortion Drone' to Fly Pills across Border into Poland," *The Guardian*, June 24, 2015, https://www.theguardian.com/world/2015/jun/24/abortion-drone-border-poland-germany-women-on-waves.

16. Stanley White, "Japanese Man Arrested for Landing Drone on PM's Office in Nuclear Protest," *Reuters*, April 24, 2015, http://www.reuters.com/article/us-japan-nuclear-drone-idUSKBN0NG04520150425.

17. Ibid.

18. This section draws heavily on research done by AlHakam Saar and published together with myself and AlHakam Shaar; see Austin Choi-Fitzpatrick and AlHakam Shaar, "Struggle over Aleppo's Story Takes to the Skies How Citizen Journalists Used Drone Technology to Challenge the Assad Regime's Narrative of Aleppo's Plight," *Al Jazeera*, February 22, 2017.

19. Michael Kimmelman, "Berlin, 1945; Grozny, 2000; Aleppo, 2016," *New York Times*, October 14, 2016, https://www.nytimes.com/2016/10/15/world/middleeast/aleppo-destruction-drone-video.html.

20. "Drone Footage Captures Devastation of East Aleppo," RT Ruptly, December 13, 2016, YouTube, https://www.youtube.com/watch?v=FH6xRh6K7-4.

21. Ibid.; "Drone Footage Shows Fierce Clashes between Syrian Army & US Backed Islamic Terrorists," TopSecret Conspiremos, December 9, 2015, YouTube, https://www.youtube.com/watch?v=IjcLEO7dcgQ.

22. Maureen Cavanaugh and Michael Lipkin, "San Diego Zoo, Northrop Grumman Developing Drones to Study Polar Bears," KPBS, November 29, 2016, http://www.kpbs.org/news/2016/nov/29/san-diego-zoo-northrop-grumman-developing-drones-s.

23. Tia Ghose, "Tiny Drones Capture Gorgeous Views of Sizzling Lava Lake," LiveScience, February 4, 2015, http://www.livescience.com/49699-drone-videos-lava-lake.html.

24. Megan Gannon, "Drone Images Reveal Buried Ancient Village in New Mexico," LiveScience, April 8, 2014, http://www.livescience.com/44679-drone-images-reveal-buried-archaeological-ruins.html.

25. Emma Marris, "Drones in Science: Fly, and Bring Me Data," *Nature*, June 12, 2013, http://www.nature.com/news/drones-in-science-fly-and-bring-me-data-1.13161.

26. "FAA Grants Drone Access to Texas and Oklahoma Panhandles for Weather Research," University of Colorado Boulder News Center, May 27, 2015, http://www.colorado.edu/news/releases/2015/05/27/faa-grants-drone-access-texas-and-oklahoma-panhandles-weather-research.

27. Jason Goldman, "Flying High: The Ethics of Monitoring Birds with Drones," *Conservation Magazine*, February 18, 2015, http://conservationmagazine.org/2015/02/flying-high-the-ethics-of-monitoring-birds-with-drones.

28. Kevin Bales, *Disposable People: New Slavery in the Global Economy* (Berkeley: University of California Press, 2012).

29. Alison Brysk and Austin Choi-Fitzpatrick, *From Human Trafficking to Human Rights: Reframing Contemporary Slavery* (Philadelphia: University of Pennsylvania Press, 2012).

30. Choi-Fitzpatrick, *What Slaveholders Think*.

31. Bales, *Disposable People*.

32. "Eyes in the Sky," Ending Slavery: Strategies for Contemporary Global Abolition, University of Nottingham online course, https://www.futurelearn.com /courses/slavery/0/steps/24401; Robin McDowell, Martha Mendoza and Margie Mason, "AP Tracks Slave Boats to Papua New Guinea," Associated Press, July 27, 2015, https://www.ap.org/explore/seafood-from-slaves/ap -tracks-slave-boats-to-papua-new-guinea.html; Kathleen Caulderwood, "Satellites Are Helping Root Out Modern Slavery in Northern India," August 7, 2017, https://news.vice.com/story/satellites-are-helping-root-out-modern -slavery-in-northern-india; Perry Chiaramonte, Fox News, October 22, 2015, "Help from Above: Satellites, Crowd Sourcing to Fight Child Slavery in West Africa," http://www.foxnews.com/world/2015/10/22/campaign-uses -satellite-images-crowd-sourcing-to-fight-child-slavery-in-west.html; Sarah Scoles, "Researchers Spy Signs of Slavery from Space," *Science* 363, no. 6429 (2019): 804, DOI: 10.1126/science.363.6429.804.

33. Quoted in Matthew Stock, "Brick by Brick: Satellite Images Could Identify Slave Labor in India," Reuters, August 3, 2017, https://www.reuters .com/article/us-india-slavery-satellites/brick-by-brick-satellite-images-could -identify-slave-labor-in-india-idUSKBN1AJ1OY.

34. Margaret E. Keck and Kathryn Sikkink, *Activists Beyond Borders: Advocacy Networks in International Politics* (Cornell University Press, 2014); Thomas Risse, Stephen C. Ropp, and Kathryn Sikkink, *The Power of Human Rights: International Norms and Domestic Change* (New York: Cambridge University Press, 1999); Thomas Risse et al., *The Persistent Power of Human Rights: From Commitment to Compliance* (New York: Cambridge University Press, 2013).

35. Martin Bulmer, Kevin Bales, and Kathryn Kish Sklar, *The Social Survey in Historical Perspective, 1880–1940* (New York: Cambridge University Press, 1991).

CHAPTER 4

1. This chapter, presented here by permission from the publisher, draws heavily on Austin Choi-Fitzpatrick, "Drones, Camera Innovations and Conceptions of Human Rights," in *Visual Imagery and Human Rights Practice*, ed. Sandra Ristovska and Monroe Price (New York: Palgrave Macmillan, 2018), 35–66.

2. Sandra Ristovska, "Witnessing and the Failure of Communication," *The Communication Review* 17, no. 2 (2014): 347–360.

3. Ibid.

4. Personal communication, May 21, 2015.

5. Sharon Sliwinski, "The Childhood of Human Rights: The Kodak on the Congo," *Journal of Visual Culture* 5, no. 3 (2006): 333–363.

6. Sam Gregory, "Cameras Everywhere: Ubiquitous Video Documentation of Human Rights, New Forms of Video Advocacy, and Considerations of Safety, Security, Dignity and Consent," *Journal of Human Rights Practice* 2, no. 2 (2010): 191–207.

7. Sharon Sliwinski, *Human Rights in Camera* (Chicago: University of Chicago Press, 2011).

8. Antigoni Memou, *Photography and Social Movements: From the Globalisation of the Movement (1968) to the Movement Against Globalisation (2001)* (Manchester, UK: Manchester University Press, 2015).

9. P. Sorokowski, A. Sorokowska, A. Oleszkiewicz, T. Frackowiak, A. Huk, and K. Pisanski, "Selfie Posting Behaviors Are Associated with Narcissism Among Men," *Personality and Individual Differences* 85 (2015): 123–127.

10. Ariella Aïsha Azoulay, *Civil Imagination: A Political Ontology of Photography* (New York: Verso Books, 2015).

11. Beaumont Newhall, *Airborne Camera; the World from the Air and Outer Space* (New York: Hastings House, 1969).

12. Susie Linfield, *The Cruel Radiance: Photography and Political Violence* (Chicago: University of Chicago Press, 2011).

13. Roland Barthes, *Camera Lucida: Reflections on Photography* (New York: Macmillan, 1981), 15, emphasis added.

14. Ibid., 106.

15. Ibid., 140.

16. Ibid., 139.

17. Ibid., 10.

18. Ibid., 10.

19. Sliwinski is not alone. Across a wave of provocative interventions, Ariel Azoulay asks, *what is photography?* (2010). Fewer ask *what is the camera?* So much work has gone into reading the image that we have generally overlooked the device. The camera is never in focus, so to speak. Exceptions exist, but for change-oriented social actors, the coin of the realm is public sentiment, witness, and the mobilization potential of the image. The image is summoned for its ability to arouse passion, frame issues, and mobilize action. For many of us, socio-political history can be recalled in a series of images of conflict, poverty, and violence, or a riveting admixture of the same. Rights violations, rights holders, claimants, struggles, abuses, and restitution are folded into a series of the *image*.

20. This is not to say that cameras have been completely ignored as cultural and political objects. Several notable exceptions exist, including Peter Wollheim, who located the camera, rather than the photograph, at the center of his inquiry. It is a pity that Peter Wollheim's masters and doctoral theses never made it to press. Defended in 1990, Wollheim's dissertation is the product of an earlier era of communication theory, but his emphasis on the technical forces behind the relations of cultural production still hold valuable lessons (Peter Wollheim, "Towards a Critical History of the 35 mm Still Photographic Camera in North America, 1896 to 1980," thesis, University of California, Los Angeles, 1993, 14). It is the 35 mm camera, Wollheim argued, that was "the visual counterpart of the guitar for the youth culture of the 60s and 70s," revolutionizing cover art and the visual aesthetic of *Rolling Stone* magazine. The punch came from the handheld camera's ability to deliver "candid, behind-the-scenes, off-stage, diaristic visual styles" (ibid.). Wollheim did not set out to explain the development of the camera vis-a-vis any particular type of use (i.e., advocacy and awareness), but he does advance a critical history of the 35 mm camera that suggests, "at least in terms of advanced industrial societies, technological innovation is inseparable from— but not necessarily synonymous with—a set of social needs" (ibid., 317). These needs, Wollheim suggests, are at "once scientific, artistic, technological,

political, military, and social" (ibid.). I very much hope Wollheim would have found this chapter worthwhile.

21. Ariella Azoulay, *The Civil Contract of Photography*, trans. Reia Mazali and Ruvik Danielli (New York: Zone, 2008), 23, 178.

22. Ibid., 24.

23. Ariella Azoulay, *Aïm Deüelle Lüski and Horizontal Photography*, vol. 16 (Leuven: Leuven University Press, 2014).

24. Ibid., 27.

25. Ariella Azoulay, "Aïm Deüelle Lüski Pinhole Cameras Philosophy & Photography," YouTube, https://www.youtube.com/watch?v=_-KuWbwGJuk&feature=youtu.be.

26. Azoulay, *Aïm Deüelle Lüski and Horizontal Photography*, 16.27.

27. Ibid., cover material.

28. Sliwinski, *Human Rights in Camera*, 58; Sliwinski, "The Childhood of Human Rights."

29. Ella McPherson, "Digital Human Rights Reporting by Civilian Witnesses: Surmounting the Verification Barrier," in *Produsing Theory in a Digital World 2.0: The Intersection of Audiences and Production in Contemporary Theory*, ed. Rebecca Ann Lind (New York: Peter Lang, 2015), 193.

30. Gillian Rose, *Visual Methodologies: An Introduction to Researching with Visual Materials* (Thousand Oaks, CA: Sage, 2016).

31. Sandra Ristovska, "The Rise of Eyewitness Video and Its Implications for Human Rights: Conceptual and Methodological Approaches," *Journal of Human Rights* 15, no. 3 (2016): 347.

32. Kaja Silverman, *The Miracle of Analogy, or the History of Photography* (Stanford, CA: Stanford University Press, 2015), 14.

33. Jonathan Crary, *Techniques of the Observer: On Vision and Modernity in the Nineteenth Century* (Cambridge, MA: MIT Press, 1992).

34. Silverman, *The Miracle of Analogy*, 15.

35. Sarah Kofman, *Camera Obscura: Of Ideology* (Cornell University Press, 1999).

36. For example, Bob Zeller, *The Blue and Gray in Black and White: A History of Civil War Photography* (Westport, CT: Greenwood Publishing Group, 2005).

37. Lawrence Douglas, *The Memory of Judgement: Making Law and History in the Trials of Holocaust* (New Haven, CT: Yale University Press, 2005), 28.

38. Silverman, *The Miracle of Analogy*.

39. Ibid.

40. Vicki Goldberg, *The Power of Photography: How Photographs Changed Our Lives* (New York: Abbeville Press, 1991).

41. Ariella Aïsha Azoulay, *Civil Imagination: A Political Ontology of Photography* (New York: Verso Books, 2015).

42. Lynn Hunt, *Inventing Human Rights: A History*, 1st ed. (New York: W.W. Norton & Co., 2007).

43. Douglas, *The Memory of Judgement*, 28.

44. Sliwinski, "The Childhood of Human Rights," 335.

45. Ibid., p. 346.

46. Sliwinski, *Human Rights in Camera*.

47. Linfield, *The Cruel Radiance*; Mark Sealy, Roger Malbert, and Alice Lobb, *Documenting Disposable People: Contemporary Global Slavery* (London: Hayward Gallery, 2008); Heide Fehrenbach and Davide Rodogno, *Humanitarian Photography* (New York: Cambridge University Press, 2015).

48. Haraway, *Simians, Cyborgs, and Women*; Rosi Braidotti, "Posthuman, All Too Human: Towards a New Process Ontology," *Theory, Culture & Society* 23, no. 7–8 (2006): 197–208; Bostrom, *Superintelligence*; Hanson, *The Age of Em*.

49. Lisa Parks, "Satellite Views of Srebrenica: Tele-Visuality and the Politics of Witnessing," *Social Identities* 7, no. 4 (2001), 589.

50. Simone Browne, *Dark Matters: On the Surveillance of Blackness* (Durham, NC: Duke University Press, 2015), 9.

51. Ibid.

52. James C. Scott, *The Art of Not Being Governed: An Anarchist History of Upland Southeast Asia* (New Haven, CT: Yale University Press, 2014).

53. Silverman, *The Miracle of Analogy*, 1.

54. But even here the drone heralds something different, since CCTVs and satellites have been fitted onto walls and into orbit *by humans*. Future systems will

rely on decisions made by complex networks of sensors and algorithms rather than fixed orbits and wall mounts.

55. Who is the photographer when a drone sets off to document a data event (perhaps Google reports that a hundred Android devices have experienced a temperature spike on the top floor of an office complex) and then sends digital footage back to a server for later review?

56. Rahul Garg and Neal Wadhwa, "Learning to Predict Depth on the Pixel 3 Phones," Google AI Blog, November 29, 2018, https://ai.googleblog.com /2018/11/learning-to-predict-depth-on-pixel-3.html; Marc Levoy and Yael Pritch, "Night Sight: Seeing in the Dark on Pixel Phones," Google AI Blog, November 14, 2018, https://ai.googleblog.com/2018/11/night-sight-seeing -in-dark-on-pixel.html.

57. "Interview with the Google Pixel 3 Camera Team," DPReview, October 11, 2018, YouTube, https://www.youtube.com/watch?v=W-bJg2L2HxA.

CHAPTER 5

1. Thanks to Nina Williams, Patrick Schoettmer, Jennifer Carter Barnett, and Dana Alan for help in compiling these examples.

2. A thousand thanks to all the friends on Facebook who helped crowdsource these references.

3. Neil Fligstein and Doug McAdam, "Putting Values and Institutions Back into the Theory of Strategic Action Fields: Response to Goldstone and Useem," *Sociological Theory* 30, no. 1 (2012): 48–50.

4. Symbolic because the ability to affect this impression is achieved through hundreds of years of colonial expropriation. The appearance that culture matters more than capital is the product of and rooted in forces both older (colonialism) and broader; see Thomas Piketty, "About Capital in the Twenty-First Century," *American Economic Review* 105, no. 5 (2015): 48–53.

5. *Why does the sky hold such high cultural value?* a bright graduate student scribbled into the margin of this manuscript. This is a wonderful question, and a topic worthy of future exploration.

6. But I'm an academic; we soldier on.

7. Winner, *Autonomous Technology,* 103.

8. James C. Scott, *Two Cheers for Anarchism* (Princeton, NJ: Princeton University Press, 2012).

9. Jennifer Schuessler, "Professor Who Learns from Peasants," *New York Times*, December 4, 2012, https://www.nytimes.com/2012/12/05/books/james-c-scott-farmer-and-scholar-of-anarchism.html.

10. Roberto Mangabeira Unger, *The Self Awakened: Pragmatism Unbound* (Cambridge, MA: Harvard University Press, 2007).

11. Owen, *Disruptive Power*, 54.

12. John Rawls, *A Theory of Justice* (Cambridge, MA: Harvard University Press, 2009), 320.

13. Worms Against Nuclear Killers was designed so that NASA and Energy Department login screens read, in part: "WORMS AGAINST NUCLEAR KILLERS ... Your System Has Been Officially WANKed." These are only two of many instances documented by Owen.

14. He continues, "Like acts of offline civil disobedience, digital efforts are ethically motivated; reject violence, a profit motive, and destruction of property; and participants accept personal responsibility for their actions" (Owen, *Disruptive Power*, 56).

15. David Uberti, "Drone Makers Struggle for Acceptance," *Boston Globe*, April 7, 2013, https://www.bostonglobe.com/business/2013/04/06/massachusetts-national-drone-companies-are-struggling-gain-public-acceptance-face-controversy/qtCg0CxAIUfrW7applrKWL/story.html.

16. The terrorist operation ISIL has strapped bombs to several devices constructed out of corrugated plastic and duct tape.

17. John Beck, "ISIL Ramps Up Fight with Weaponised Drones," *Al Jazeera*, January 3, 2017, https://www.aljazeera.com/indepth/features/2016/12/isil-ramps-fight-weaponised-drones-161231130818470.html.

18. Martin Dodge and Rob Kitchin, *Atlas of Cyberspace* (London: Addison-Wesley, 2001), xi.

19. Uri Gordon, "Anarchism and the Politics of Technology," *WorkingUSA* 12, no. 3 (2009): 489–503; Giorel Curran and Morgan Gibson, "Wikileaks, Anarchism and Technologies of Dissent," *Antipode* 45, no. 2 (2013): 298.

20. Murray Bookchin, *The Ecology of Freedom: The Emergence and Dissolution of Hierarchy* (Palo Alto, CA: Cheshire Books, 1982).

21. Curran and Gibson, "Wikileaks, Anarchism and Technologies of Dissent," 299.

22. I have Patrick Meier to thank for this observation.

23. Arthur Holland Michel and Dan Gettinger, "Drone Year in Review" (Annandale-on-Hudson, NY: Center for the Study of the Drone at Bard College, 2017).

24. "Syracuse Is Fifth City to Pass Anti-Drone Resolution," WarIsACrime.org, accessed December 16, 2015.

25. "Resolution on Drone Aircraft," *City of Northampton,* July 11, 2013, accessed December 16, 2015.

26. Winner, *Autonomous Technology,* 88.

27. Hannah Arendt, *The Human Condition* (Chicago: University of Chicago Press, 2013), 190, quoted in Winner, *Autonomous Technology,* 88.

28. Jaron Lanier, *Who Owns the Future?* (New York: Simon & Schuster, 2014).

29. Arthur Holland Michel, "Counter-Drone Systems" (Annandale-on-Hudson, NY: Center for the Study of the Drone at Bard College, 2018).

30. Droneshield company website, https://www.droneshield.com.

31. Charles Arthur, "SkyGrabber: The $26 Software Used by Insurgents to Hack into US Drones," *The Guardian,* December 17, 2009, https://www.theguardian.com/technology/2009/dec/17/skygrabber-software-drones-hacked.

32. Virus-copter, GitHub, https://github.com/substack/virus-copter; Andrew Tarantola, "This Virus-Copter Is a Digital Typhoid Mary," *Gizmodo,* December 10, 2012, http://gizmodo.com/5967209/this-virus-copter-is-a-digital-typhoid-mary.

33. "Counter-UAS Technologies," Batelle company website, https://www.battelle.org/government-offerings/national-security/tactical-systems-vehicles/tactical-equipment/counter-UAS-technologies.

34. Andrew Moseman, "This Drone Interceptor Captures Your Pathetic Puny Drone With a Net," *Popular Mechanics*, February 11, 2015, http://www

.popularmechanics.com/flight/drones/a14032/france-dispatches-a-net
-carrying-bully-drone-to-catch.

35. Tom Brant, "AI Program Wins Dogfight Against USAF Fighter Pilot," June 29, 2016, http://in.pcmag.com/drones/104754/news/ai-program-wins-dogfight -against-usaf-fighter-pilot.

36. Anna Giaritelli, "Judge Sides with Man Who Shot Down Drone," *Washington Examiner*, October 30, 2015, http://www.washingtonexaminer.com /judge-sides-with-man-who-shot-down-drone/article/2575323.

37. Chris Matyszczyk, "Drone Shooter Pleads Guilty," *CNET*, February 14, 2016, https://www.cnet.com/news/man-who-shot-down-drone-pleads-guilty/.

38. Steve Dent, "A Russian Drone Hunts Other Drones with a Shotgun," *Endgadget*, April 1, 2019, https://www.engadget.com/2019/04/01/russian -shotgun-packing-drone/.

39. Yamada, Takayuki, Seiichi Gohshi, and Isao Echizen. "Privacy Visor: Method Based on Light Absorbing and Reflecting Properties for Preventing Face Image Detection," in *IEEE International Conference on Systems, Man, and Cybernetics* (Washington, DC: IEEE Computer Society, 2013), 1572–1577.

40. Mrn P. Kadaba, H. K. Ramakrishnan, and M. E. Wootten, "Measurement of Lower Extremity Kinematics During Level Walking," *Journal of Orthopaedic Research* 8, no. 3 (1990): 383–392.

41. James Hawes, "Hated in the Nation," in *Black Mirror*, ed. Charlie Brooker (United Kingdom: BBC, 2016).

42. Charlie Brooker, "White Christmas," in *Black Mirror*, ed. Charlie Brooker (United Kingdom: BBC, 2014).

43. Henrike Schmidt, "From a Bird's Eye Perspective: Aerial Drone Photography and Political Protest. A Case Study of the Bulgarian# Resign Movement 2013," *Digital Icons: Studies in Russian, Eurasian and Central European New Media* 13 (2015): 1–27.

44. Cathy O'Neil, *Weapons of Math Destruction: How Big Data Increases Inequality and Threatens Democracy* (New York: Broadway Books, 2017).

45. Dan Gettinger, "Public Safety Drones: An Update," Center for the Study of the Drone at Bard College, May 28, 2018, https://dronecenter.bard.edu /public-safety-drones-update.

46. Dronefly company website, "Police Drone Infographic," https://www.dronefly.com/pages/police-drone-infographic.

47. In New York City, public space that intersects with private access points is regularly colonized by private interests. See Elis Rosenberg's "A 'Members Only' Public Space in Manhattan? Join the Club," *New York Times,* April 19, 2017, https://www.nytimes.com/2017/04/19/nyregion/public-space-trump-tower.html.

48. MacLeod, personal interview.

49. Vestergaard, personal correspondence.

50. Jesper explains how he became interested in drone graffiti: "Graffiti artists walk inside the building at night, use broomsticks to paint rooftops, so they can be seen from far away. I saw one that said *Give Up, Drop Everything.* It was nice—there was some space missing between the letters, as if some of the text had been dropped. This was in late 2013—I was wondering how it got up there, so I started investigating. I'm afraid of heights ... so I started exploring how technology could be used to do this" (personal communication).

51. Here we can imagine a two by two, in which it is possible to inquire about application and visibility. Public art can be applied by humans or by drone, and can by seen by humans (on the ground) or only by drone. Presumably, therefore, some art can be only made and seen by drones (or via drones). I have Carolyn Ross to thank for this observation.

52. Jim Woodman, *Nazca: Journey to the Sun* (New York: Pocket, 1977).

53. Graham, *Vertical,* 50.

54. Parks, *Cultures in Orbit: Satellites and the Televisual,* 13.

55. Ibid., 171.

56. Philipp Kaiser and Miwon Kwon, *Ends of the Earth: Land Art to 1974* (New York: Prestel Pub, 2012).

57. Jeffrey Kastner and Brian Wallis, *Land and Environmental Art* (London: Phaidon Press, 1998).

58. George Thomas Baker and Lynne Cooke, *Robert Smithson: Spiral Jetty: True Fictions, False Realities* (Berkeley: University of California Press, 2005).

59. Kaiser and Kwon, *Ends of the Earth*; Robert Smithson, "Towards the Development of an Air Terminal Site," *Artforum* 6, no. 10 (1967): 52–60; Robert

Smithson, *Robert Smithson: The Collected Writings* (Berkeley: University of California Press, 1996).

60. Kastner and Wallis, *Land and Environmental Art.*

61. Robert Morris, "Aligned with Nazca," *Artforum* 14, no. 2 (1975): 26–39, quoted in Kastner and Wallis, *Land and Environmental Art,* 30.

CHAPTER 6

1. Dana DiFilippo, "Bucks Gun Club Accused of Abusing Pigeons," *The Philadelphia Inquirer,* February 11, 2016, http://www.philly.com/philly/news/20160212_Bucks_Gun_club_investigated_for_alleged_animal_cruelty.html.

2. Sarah A. Soule, "The Student Divestment Movement in the United States and Tactical Diffusion: The Shantytown Protest," *Social Forces* 75, no. 3 (1997): 855–882; Verta Taylor and Nella Van Dyke, "'Get up, Stand Up': Tactical Repertoires of Social Movements," in *The Blackwell Companion to Social Movements,* ed. Hanspeter Kriesi, Sarah A. Soule, and David A. Snow (Malden, MA: Blackwell, 2004), 262–293; Dan J. Wang and Sarah A. Soule, "Social Movement Organizational Collaboration: Networks of Learning and the Diffusion of Protest Tactics, 1960–1995," *American Journal of Sociology* 117, no. 6 (2012): 1674–1722.

3. Philip N. Howard, *The Digital Origins of Dictatorship and Democracy: Information Technology and Political Islam* (New York: Oxford University Press, 2010); Earl and Kimport, *Digitally Enabled Social Change*; Bennett and Segerberg, *The Logic of Connective Action*; Bruce A. Bimber, Andrew J. Flanagin, and Cynthia Stohl, *Collective Action in Organizations: Interaction and Engagement in an Era of Technological Change* (New York: Cambridge University Press, 2012); Chadwick, *The Hybrid Media System*; Tufekci, *Twitter and Tear Gas*; Owen, *Disruptive Power*; Evgeny Morozov, *The Net Delusion: The Dark Side of Internet Freedom* (New York: Public Affairs, 2011); Morozov, *To Save Everything, Click Here.*

4. Jeremy Anderson, "Moral Problems of Remote Sensing Technology," *Antipode* 1, no. 1 (1969): 54.

5. Jennifer Gabrys, *Program Earth: Environmental Sensing Technology and the Making of a Computational Planet,* (Minneapolis: University of Minnesota Press, 2016).

6. Boyd, Doreen S., et al. "Slavery from Space: Demonstrating the Role for Satellite Remote Sensing to Inform Evidence-Based Action Related to UN SDG Number 8," *ISPRS Journal of Photogrammetry and Remote Sensing* 142 (2018): 380–388.

7. Parks and Starosielski, *Signal Traffic*; Starosielski, *The Undersea Network*.

8. Jessie Hempel, "Inside Facebook's Ambitious Plan to Connect the Whole World," *Wired* (2016).

9. Benjamin Peters, *How Not to Network a Nation: The Uneasy History of the Soviet Internet* (Cambridge, MA: MIT Press, 2016).

10. Xiao Qiang, "The Battle for the Chinese Internet," *Journal of Democracy* 22, no. 2 (2011): 47–61.

11. Philip N. Howard and Muzammil M. Hussain, *Democracy's Fourth Wave?: Digital Media and the Arab Spring* (New York: Oxford University Press, 2013).

12. Beyer, *Expect Us*.

13. Ibid.

14. Steven Livingston and Douglas A. van Belle, "The Effects of Satellite Technology on Newsgathering from Remote Locations," *Political Communication* 22, no. 1 (2005): 45–62.

15. See Nan Lin, *The Struggle for Tiananmen: Anatomy of the 1989 Mass Movement* (Westport, CT: Praeger Publishers, 1992); Zhou He, *Mass Media and Tiananmen Square* (New York: Nova Science Publishers, 1996); and Sid Pike, *We Changed the World: Memoirs of a CNN Global Satellite Pioneer* (St. Paul, MN: Paragon House, 2005). As prices fall, satellites are increasingly used to monitor human rights violations, surveil prison camps (North Korea), and identify mass graves (Burundi), though this usage might be better thought of as *input*, since the satellites in question are data-collecting tools.

16. Tom Rosentiel, "Public Attitudes Toward the War in Iraq: 2003–2008," Pew Research Center, March 19, 2008, http://www.pewresearch.org/2008/03/19/public-attitudes-toward-the-war-in-iraq-20032008/

17. In my family we would say that something immovable *didn't budge*, but I learned from my young friend Hammer that the word can also be used in a positive construction, as in: "We budged the bully from the pavement."

18. I have the maven John Holland to thank for the equivalencies here. I vetoed his original idea, which was to convert everything into the VW unit of analysis. For anyone interested, once this book went to press, it was .05 SVWU (standard VW unit), or, roughly .13 stone.

19. New developments in predictive computing and broader inputs from data arrays are likely to increase the power of future analysis, but the tools are likely to remain the same.

20. Bostrom, *Superintelligence*.

21. Gabrys, *Program Earth*, 49.

22. Gina Neff and Dawn Nafus, *The Quantified Self* (Cambridge, MA: MIT Press, 2016).

23. The San Diego Zoo's official title is San Diego Zoo Global.

24. McCarthy and Zald, "Resource Mobilization and Social Movements."

25. Robert Michels, *Political Parties: A Sociological Study of the Oligarchical Tendencies of Modern Democracy* (New York: Hearst's International Library Company, 1915).

26. Darcy K. Leach, "The Iron Law of What Again? Conceptualizing Oligarchy across Organizational Forms," *Sociological Theory* 43 (2005): 312–337, 313.

27. Piven and Cloward, *Poor People's Movements*, 697.

28. Michael P. Young, *Bearing Witness against Sin: The Evangelical Birth of the American Social Movement* (Chicago: University of Chicago Press, 2006).

29. Seymour Drescher, *Capitalism and Antislavery: British Mobilization in Comparative Perspective* (New York: Oxford University Press, 1987), 221, fn. 70.

30. John R. Oldfield, *Popular Politics and British Anti-Slavery: The Mobilisation of Public Opinion against the Slave Trade 1787–1807* (New York: Routledge, 2012), 1.

31. Drescher, *Capitalism and Antislavery*, 86.

32. Ibid., 92.

33. Oldfield, *Popular Politics and British Anti-Slavery*, 114.

34. Ethan A. Nadelmann, "Global Prohibition Regimes: The Evolution of Norms in International Society," *International Organization* 44, no. 4 (1990): 479–

526; Adam Hochschild, *Bury the Chains: Prophets and Rebels in the Fight to Free an Empire's Slaves* (New York: Houghton Mifflin Harcourt, 2006).

35. Tilly, *Regimes and Repertoires*, 56.

36. A. Outwater, N. Abrahams, and J. C. Campbell, "Women in South Africa: Intentional Violence and HIV/AIDS: Intersections and Prevention," *Journal of Black Studies* 35, no. 135 (2005): 135–154; A. M. Fox, S. S. Jackson, N. B. Hansen, N. Gasa, M. Crewe, and K. J. Sikkema, "In Their Own Voices: A Qualitative Study of Women's Risk for Intimate Partner Violence and HIV in South Africa," *Violence Against Women* 13, no. 58 (2007): 583–602.

37. Chadwick, *The Hybrid Media System*.

38. Charles Tilly, "From Interactions to Outcomes in Social Movements," in *How Social Movements Matter*, ed. Marco Giugni, Doug McAdam, and Charles Tilly (Minneapolis: University of Minnesota Press, 1999), 253–270; Charles Tilly and Lesley J. Wood, *Social Movements 1768–2012* (New York: Routledge, 2015).

39. Biggs, "Size Matters"; Austin Choi-Fitzpatrick, Tautvydas Juskauskas, and Md Boby Sabur, "All the Protestors Fit to Count: Using Geospatial Affordances to Estimate Protest Event Size," *Interface* 10 (2018): 273–281.

40. As an aside, as it enters the third decade of its existence, the AIDS Memorial Quilt is now 54 tons.

41. McDonnell, *Best Laid Plans*; Terence E. McDonnell, "Cultural Objects as Objects: Materiality, Urban Space, and the Interpretation of Aids Campaigns in Accra, Ghana," *American Journal of Sociology* 115, no. 6 (2010): 1800–1852.

42. Joseph E. Luders, *The Civil Rights Movement and the Logic of Social Change* (New York: Cambridge University Press, 2010).

43. Brayden G. King and Sarah A. Soule, "Social Movements as Extra-Institutional Entrepreneurs: The Effect of Protests on Stock Price Returns," *Administrative Science Quarterly* 52, no. 3 (2007): 413–442.

44. Mark Traugott, *The Insurgent Barricade* (Berkeley: University of California Press, 2010).

45. Steve Jones, *Revolutionary Science: Transformation and Turmoil in the Age of the Guillotine* (New York: Pegasus, 2017), 341.

46. The history of the urinal, like so many modest human accomplishments, is difficult to trace. It was patented in the United States in 1866 by one Andrew Rankin, but of course variants were in use for far longer.

47. Scott, *The Art of Not Being Governed.*

48. Nagy and Neff, "Imagined Affordance."

49. Donald A. MacKenzie, *Knowing Machines: Essays on Technical Change* (Cambridge, MA: MIT Press, 1998).

50. Ozge Dilaver, "Making Sense of Innovations: A Comparison of Personal Computers and Mobile Phones," *New Media & Society* 16, no. 8 (2014): 1214–1232, 1214.

51. My factors generally echo those of Rogers from a half-century earlier: relative advantage, complexity, trialability, observability, and compatibility.

52. Donald A. Norman, *The Design of Everyday Things* (New York: Basic Books, 2013), 11.

53. Elisabeth S. Clemens, "Organizational Repertoires and Institutional Change: Women's Groups and the Transformation of US Politics, 1890–1920," *American Journal of Sociology* (1993): 755–798.

54. McCarthy and Zald, "Resource Mobilization and Social Movements."

55. I am grateful to an anonymous reviewer for this observation.

56. Sliwinski, *Human Rights in Camera.*

57. Livingston and Belle, "The Effects of Satellite Technology on Newsgathering."

58. Winner, *The Whale and the Reactor,* 17.

59. Dilaver, "Making Sense of Innovations"; Burstein, "The Impact of Public Opinion on Public Policy."

60. I have Alex Hanna to thank for this observation.

61. Everett M. Rogers, *Diffusion of Innovations* (New York: Simon and Schuster, 2010).

62. Tilly, *Regimes and Repertoires,* 57.

63. Ibid., 43.

64. Thanks to Everard Kidder Meade for help here.

65. Bradshaw and Howard, "Troops, Trolls and Troublemakers"; Howard, Kollanyi, and Woolley, "Bots and Automation."

66. Chadwick, *The Hybrid Media System*.

67. Patrick Meier, for example, has for some time done the hard work of figuring out what exactly works in the UAV space and what rate of adoption is likely as we move forward.

68. Winner, *The Whale and the Reactor*, 17. In reviewing this volume, Lars Almquist asked me to define the *we* in this sentence, asking: "At what point is the line drawn between a completely free market and a totalitarian state controlling the release of technology to the masses? What's the optimal middle ground, and how do you engage that? Wouldn't a civil libertarian disagree with your trust in government to decide which technologies should be brought to market? Why would/should we limit the set of options available to us and why is the government, the state you're actively documenting protests against with your drone, the one to regulate the availability of those technologies?" While there is no simple answer to these questions, my preference is for a polity in which such issues are identified, discussed, and acted upon. The result of such debate is contingent on the case and context, an observation that might not please Lars, but is as close as I'm able to get.

69. Ibid., but see also Winner, *Autonomous Technology*, and Langdon Winner, "Do Artifacts Have Politics?" *Daedalus* (1980): 121–136.

THEORETICAL AFTERWORD 1

1. Woolley and Howard, *Computational Propaganda*; Larry Jay Diamond and Marc F. Plattner, *Liberation Technology: Social Media and the Struggle for Democracy* (Baltimore, MD: Johns Hopkins University Press, 2012).

2. Earl and Kimport, *Digitally Enabled Social Change*.

3. Donna J. Haraway, *Staying with the Trouble: Making Kin in the Chthulucene* (Durham, NC: Duke University Press, 2016).

4. Mancur Olson, *The Logic of Collective Action: Public Goods and the Theory of Groups* (Cambridge, MA: Harvard University Press, 2009).

5. Bennett and Segerberg, *The Logic of Connective Action*, cover copy.

6. Many thanks to Lars Almquist for suggesting these examples.

7. Bennett and Segerberg, *The Logic of Connective Action*, 25.

8. Brayden King, "Digital Media, Connective Action, and Social Movements," *OrgTheory* (blog), March 10, 2014, https://orgtheory.wordpress.com/2014/03/10/digital-media-connective-action-and-social-movements/press.com/2014/03/10/digital-media-connective-action-and-social-movements/.

9. Roscigno and Danaher, "The Voice of Southern Labor."

10. Rachel Schurman and William A. Munro, *Fighting for the Future of Food: Activists versus Agribusiness in the Struggle over Biotechnology* (Minneapolis: University of Minnesota Press, 2013).

11. E.g., Jennifer Earl, Donna Della Porta, Deanna Rohlinger, Mario Diani, and Dana Fischer.

12. Chadwick, *The Hybrid Media System*, 5.

13. Howard, *The Digital Origins of Dictatorship and Democracy*; Philip N. Howard and Muzammil M. Hussain, "The Role of Digital Media," *Journal of Democracy* 22, no. 3 (2011): 35–48.

14. Bennett and Segerberg, *The Logic of Connective Action*, 41.

15. Chadwick, *The Hybrid Media System*.

16. Ibid., 4.

17. Ibid., 4.

18. Ibid., 4.

19. Carolyn Marvin, *When Old Technologies Were New: Thinking About Electric Communication in the Late Nineteenth Century* (New York: Oxford University Press, 1990), 3, emphasis added.

20. Livingston and Walter-Drop, *Bits and Atoms*.

21. Jennifer Earl and R. Kelly Garrett, "The New Information Frontier: Toward a More Nuanced View of Social Movement Communication," *Social Movement Studies* 16, no. 4 (2017): 479–493.

THEORETICAL AFTERWORD 2

1. I thank John Krinsky for suggesting that I use this paragraph to foreground my argument before wading into details. He suggested how to get started, and I gratefully followed his lead.

2. Many thanks to John Krinsky for this observation.

3. Andrew G. Walder, "Political Sociology and Social Movements," *Annual Review of Sociology* 35 (2009): 393–412, 393.

4. Rory McVeigh, *The Rise of the Ku Klux Klan: Right-Wing Movements and National Politics* (Minneapolis: University of Minnesota Press, 2009); Michael Schwartz, *Radical Protest and Social Structure: The Southern Farmers' Alliance and Cotton Tenancy, 1880–1890* (Chicago: University of Chicago Press, 1988).

5. Gabriel Hetland and Jeff Goodwin, "The Strange Disappearance of Capitalism from Social Movement Studies," in *Marxism and Social Movements*, ed. Colin Barker, Laurence Cox, John Krinsky, and Alf Gunvald Nilsen (Leiden, NL: Brill, 2013), 82–102.

6. Immanuel Wallerstein, *The Modern World-System I: Capitalist Agriculture and the Origins of the European World-Economy in the Sixteenth Century* (Berkeley: University of California Press, 1974).

7. Schwartz, *Radical Protest and Social Structure*.

8. Tilly, *Regimes and Repertoires*, 42.

9. Ibid., 57.

10. Ibid., 56.

11. McCarthy and Zald, "Resource Mobilization and Social Movements."

12. Olson, *The Logic of Collective Action*, 124.

13. J. Craig Jenkins and Charles Perrow, "Insurgency of the Powerless: Farm Worker Movements (1946–1972)," *American Sociological Review* (1977): 249–268; McAdam, *Political Process and the Development of Black Insurgency*.

14. Robert D. Benford and David A. Snow, "Framing Processes and Social Movements: An Overview and Assessment," *Annual Review of Sociology* 26, no. 1 (2000): 611–639.

15. Stefania Milan, *Social Movements and Their Technologies: Wiring Social Change* (New York, NY: Palgrave Macmillan, 2013).

16. Charles Tilly, *From Mobilization to Revolution* (New York: McGraw-Hill, 1978); Herbert P. Kitschelt, "Political Opportunity Structures and Political Protest: Anti-Nuclear Movements in Four Democracies," *British Journal of Political Science* 16, no. 01 (1986): 57–85. Tarrow, *Power in Movement*; McAdam, *Political Process and the Development of Black Insurgency*.

17. McAdam, *Political Process and the Development of Black Insurgency.*

18. Nikolai D. Kondratieff, "The Long Waves in Economic Life," *Review: A Journal of the Fernand Braudel Center* 2 (1979): 519–562; Simon Kuznets, "Economic Growth and Income Inequality," *The American Economic Review* 45, no. 1 (1955): 1–28; Simon Smith Kuznets, *Secular Movements in Production and Prices* (Boston: Houghton Mifflin, 1930); Daniel Smihula, "The Waves of the Technological Innovations of the Modern Age and the Present Crisis as the End of the Wave of the Informational Technological Revolution," *Studia Politica Slovaca*, no. 1 (2009): 32–47.

19. Jeff Goodwin and James M. Jasper, "Caught in a Winding, Snarling Vine: The Structural Bias of Political Process Theory," *Sociological Forum* 14, no. 1 (1999): 27–54.

20. Choi-Fitzpatrick, *What Slaveholders Think*; and see McAdam, *Political Process and the Development of Black Insurgency.*

21. Thomas Parke Hughes, *Rescuing Prometheus: Four Monumental Projects That Changed the Modern World* (New York: Pantheon Books, 1998).

22. MacKenzie, *Knowing Machines.*

23. Karl Marx, *The 18th Brumaire of Louis Bonaparte* (Cabin John, MD: Wildside Press LLC, 2008), 15.

24. Robert King Merton and Robert C. Merton, *Social Theory and Social Structure* (New York: Simon and Schuster, 1968); Anthony Giddens, *New Rules of Sociological Method: A Positive Critique of Interpretative Sociologies* (New York: John Wiley & Sons, 2013) and especially *Central Problems in Social Theory* (Berkeley: University of California Press, 1979), 49–95; Ann Swidler, "Culture in Action: Symbols and Strategies," *American Sociological Review* (1986): 273–286; Mustafa Emirbayer and Jeff Goodwin, "Network Analysis, Culture, and the Problem of Agency," *American Journal of Sociology* 99, no. 6 (1994): 1411–1454.

25. Andrew G. Walder, "Political Sociology and Social Movements," *Annual Review of Sociology* 35 (2009): 393–412.

BIBLIOGRAPHY

Adey, Peter. *Aerial Life: Spaces, Mobilities, Affects.* Hoboken, NJ: John Wiley & Sons, 2010.

Anderson, Benedict. *Imagined Communities: Reflections on the Origin and Spread of Nationalism.* New York: Verso Books, 2006.

Anderson, Jeremy. "Moral Problems of Remote Sensing Technology." *Antipode* 1, no. 1 (1969): 54–57.

Arendt, Hannah. *The Human Condition.* Chicago: University of Chicago Press, 2013.

Azoulay, Ariella Aïsha. *Aïm Deüelle Lüski and Horizontal Photography.* Leuven, NL: Leuven University Press, 2014.

Azoulay, Ariella Aïsha. *The Civil Contract of Photography.* Trans. Reia Mazali and Ruvik Danielli. New York: Zone, 2008.

Azoulay, Ariella Aïsha. *Civil Imagination: A Political Ontology of Photography.* New York: Verso Books, 2015.

Azoulay, Ariella Aïsha. "What Is a Photograph? What Is Photography?" *Philosophy of Photography* 1, no. 1 (2010): 9–13.

Baker, George Thomas, and Lynne Cooke. *Robert Smithson: Spiral Jetty: True Fictions, False Realities.* Berkeley: University of California Press, 2005.

Baker, William John. *A History of the Marconi Company 1874–1965.* New York: Routledge, 2013.

Bales, Kevin. *Disposable People: New Slavery in the Global Economy.* Berkeley: University of California Press, 2012.

Barthes, Roland. *Camera Lucida: Reflections on Photography.* New York: Macmillan, 1981.

Bar-Yam, Yaneer. *Dynamics of Complex Systems.* New York: Routledge, 2018.

Bélanger, Pierre. "Altitudes of Urbanization." *Tunnelling and Underground Space Technology Incorporating Trenchless Technology Research*, no. 55 (2016): 5–7.

Benford, Robert D., and David A. Snow. "Framing Processes and Social Movements: An Overview and Assessment." *Annual Review of Sociology* (2000): 611–639.

Benjamin, Medea. *Drone Warfare: Killing by Remote Control.* New York: Verso Books, 2013.

Bennett, Jane, Pheng Cheah, Melissa A. Orlie, Elizabeth Grosz, Diana Coole, and Samantha Frost. *New Materialisms: Ontology, Agency, and Politics.* Durham, NC: Duke University Press, 2010.

Bennett, W. Lance and Alexandra Segerberg. *The Logic of Connective Action: Digital Media and the Personalization of Contentious Politics.* Cambridge, UK: Cambridge University Press, 2013.

Bergen, Peter L., and Daniel Rothenberg. *Drone Wars: Transforming Conflict, Law, and Policy.* New York: Cambridge University Press, 2015.

Beyer, Jessica Lucia. *Expect Us: Online Communities and Political Mobilization.* New York: Oxford University Press, 2014.

Biggs, Michael. "Size Matters: Quantifying Protest by Counting Participants." *Sociological Methods & Research* 47, no. 3 (2016): 351–383.

Bijker, Wiebe E., Thomas P. Hughes, Trevor Pinch, and Deborah G. Douglas. *The Social Construction of Technological Systems: New Directions in the Sociology and History of Technology.* Cambridge, MA: MIT Press, 2012.

Bijker, Wiebe, and John Law. *Shaping Technology/Building Society: Studies in Sociotechnical Change.* Cambridge, MA: MIT Press, 1994.

Bimber, Bruce A., Andrew J. Flanagin, and Cynthia Stohl. *Collective Action in Organizations: Interaction and Engagement in an Era of Technological Change.* New York: Cambridge University Press, 2012.

Bob, Clifford. *The Global Right Wing and the Clash of World Politics.* New York: Cambridge University Press, 2012.

Boczkowski, Pablo, and Leah A. Lievrouw. "Bridging STS and Communication Studies: Scholarship on Media and Information Technologies." In *The Handbook of Science and Technology Studies*, 3rd ed., edited by Edward J. Hackett, Olga Amsterdamska, Michael Lynch, and Judy Wajcman, 949–978. Cambridge, MA: MIT Press, 2008.

Bookchin, Murray. *The Ecology of Freedom: The Emergence and Dissolution of Hierarchy.* Palo Alto, CA: Cheshire Books, 1982.

Bostrom, Nick. *Superintelligence: Paths, Dangers, Strategies.* Oxford, UK: Oxford University Press, 2014.

Bradshaw, Samantha, and Philip N. Howard. "Troops, Trolls and Troublemakers: A Global Inventory of Organized Social Media Manipulation." Working paper 2017.12. Oxford, UK: Project on Computational Propaganda, 2017.

Braidotti, Rosi. "Posthuman, All Too Human: Towards a New Process Ontology." *Theory, Culture & Society* 23, no. 7–8 (2006): 197–208.

Bromley, Lars. "Eye in the Sky: Monitoring Human Rights Abuses Using Geospatial Technology." *Georgetown Journal of International Affairs* (2009): 159–168.

Brooker, Charlie. "White Christmas." In *Black Mirror*, edited by Charlie Brooker, 74 minutes. United Kingdom: BBC, 2014.

Brown, Andrew R. "Understanding Musical Practices as Agency Networks." In *Proceedings of the Seventh International Conference on Computational Creativity.* Paris: Association of Computational Creativity, 2016.

Brysk, Alison, and Austin Choi-Fitzpatrick. *From Human Trafficking to Human Rights: Reframing Contemporary Slavery.* Philadelphia: University of Pennsylvania Press, 2012.

Bulmer, Martin, Kevin Bales, and Kathryn Kish Sklar. *The Social Survey in Historical Perspective, 1880–1940.* New York: Cambridge University Press, 1991.

Burstein, Paul. "The Impact of Public Opinion on Public Policy: A Review and an Agenda." *Political Research Quarterly* 56, no. 1 (2003): 29–40.

Chadwick, Andrew. *The Hybrid Media System: Politics and Power.* New York: Oxford University Press, 2017.

Chadwick, Nora K. "The Kite: A Study in Polynesian Tradition." *The Journal of the Royal Anthropological Institute of Great Britain and Ireland* 61 (1931): 455–491.

Chamayou, Grégoire, and Janet Lloyd. *A Theory of the Drone.* New York: The New Press, 2015.

Chenoweth, Erica, and Maria J. Stephan. *Why Civil Resistance Works: The Strategic Logic of Nonviolent Conflict.* New York: Columbia University Press, 2011.

Choi-Fitzpatrick, Austin. "Drones, Camera Innovations and Conceptions of Human Rights," in *Visual Imagery and Human Rights Practice,* edited by Sandra Ristovska and Monroe Price, 35–66. New York: Palgrave Macmillan, 2018.

Choi-Fitzpatrick, Austin. "Drones for Good: Technological Innovations, Social Movements, and the State." *Journal of International Affairs* 68, no. 1 (2014): 19.

Choi-Fitzpatrick, Austin. *What Slaveholders Think: How Contemporary Perpetrators Rationalize What They Do.* New York: Columbia University Press, 2017.

Choi-Fitzpatrick, Austin, Dana Chavarria, Elizabeth Cychosz, John Paul Dingens, Michael Duffey, Katherine Koebel, Sirisack Siriphanh, et al. *Up in the Air: A Global Estimate of Non-Violent Drone Use 2009–2015.* 2016. https://doi.org/10.22371/08 .2016.001.

Choi-Fitzpatrick, Austin, and Tautvydas Juskauskas. "Up in the Air: Applying the Jacobs Crowd Formula to Drone Imagery." *Procedia Engineering* 107 (2015): 273–281.

Choi-Fitzpatrick, Austin, Tautvydas Juskauskas, and Md Boby Sabur. "All the Protestors Fit to Count: Using Geospatial Affordances to Estimate Protest Event Size." *Interface* 10, no. 1-2 (2018): 297–321.

Choi-Fitzpatrick, Austin, and AlHakam Shaar. "Struggle over Aleppo's Story Takes to the Skies: How Citizen Journalists Used Drone Technology to Challenge the Assad Regime's Narrative of Aleppo's Plight." *Al Jazeera,* February 22, 2017.

Clemens, Elisabeth S. "Organizational Repertoires and Institutional Change: Women's Groups and the Transformation of Us Politics, 1890–1920." *American Journal of Sociology* (1993): 755–798.

Cole, Christopher, and E. F. Cheesman. *The Air Defence of Great Britain 1914–1918.* London: Putnam, 1984.

Cole, Michael, and James V. Wertsch. "Beyond the Individual-Social Antinomy in Discussions of Piaget and Vygotsky." *Human Development* 39, no. 5 (1996): 250–256.

Cortright, David, Rachel Fairhurst, and Kristen Wall. *Drones and the Future of Armed Conflict: Ethical, Legal, and Strategic Implications.* Chicago: University of Chicago Press, 2015.

Council, National Research. *People and Pixels: Linking Remote Sensing and Social Science.* National Academies Press, 1998.

Crary, Jonathan. *Techniques of the Observer: On Vision and Modernity in the Nineteenth Century.* Cambridge, MA: MIT Press, 1992.

Crossley, Alison Dahl. "Facebook Feminism: Social Media, Blogs, and New Technologies of Contemporary US Feminism." *Mobilization* 20, no. 2 (2015): 253–268.

Crouch, Tom D. *Lighter Than Air: An Illustrated History of Balloons and Airships.* Baltimore, MD: Johns Hopkins University Press, 2009.

Curran, Giorel, and Morgan Gibson. "Wikileaks, Anarchism and Technologies of Dissent." *Antipode* 45, no. 2 (2013): 294–314.

Davis, Jenny L., and James B Chouinard. "Theorizing Affordances: From Request to Refuse." *Bulletin of Science, Technology & Society* 36, no. 4 (2016): 241–248.

Diamond, Larry Jay, and Marc F. Plattner. *Liberation Technology: Social Media and the Struggle for Democracy.* Baltimore, MD: Johns Hopkins University Press, 2012.

Dilaver, Ozge. "Making Sense of Innovations: A Comparison of Personal Computers and Mobile Phones." *New Media & Society* 16, no. 8 (2014): 1214–1232.

Dodge, Martin, and Rob Kitchin. *Atlas of Cyberspace.* Vol. 158. London: Addison-Wesley, 2001.

Dosemagen, Shannon, Jeffrey Warren, and Sara Wylie. "Grassroots Mapping: Creating a Participatory Map-Making Process Centered on Discourse." *Journal of Aesthetics and Protest* 8 (2011).

Douglas, Lawrence. *The Memory of Judgment: Making Law and History in the Trials of Holocaust.* New Haven, CT: Yale University Press, 2005.

Dourish, Paul. *The Stuff of Bits: An Essay on the Materialities of Information.* Cambridge, MA: MIT Press, 2017.

Drescher, Seymour. *Capitalism and Antislavery: British Mobilization in Comparative Perspective.* New York: Oxford University Press, 1987.

Earl, Jennifer, and R. Kelly Garrett. "The New Information Frontier: Toward a More Nuanced View of Social Movement Communication." *Social Movement Studies* 16, no. 4 (2017): 479–493.

Earl, Jennifer, and Katrina Kimport. *Digitally Enabled Social Change: Activism in the Internet Age.* Cambridge, MA: MIT Press, 2011.

Edgerton, David. *The Shock of the Old: Technology and Global History since 1900.* New York: Oxford University Press, 2007.

Edwards, Michael. *Civil Society.* Cambridge, UK: Polity, 2009.

Elden, Stuart. "Secure the Volume: Vertical Geopolitics and the Depth of Power." *Political Geography* 34 (2013): 35–51.

Emirbayer, Mustafa, and Jeff Goodwin. "Network Analysis, Culture, and the Problem of Agency." *American Journal of Sociology* 99, no. 6 (1994): 1411–1454.

Emirbayer, Mustafa, and Ann Mische. "What Is Agency?" *American Journal of Sociology* 103, no. 4 (1998): 962–1023.

Evans, Charles M. *The War of the Aeronauts: A History of Ballooning During the Civil War.* Mechanicsburg, PA: Stackpole Books, 2002.

Fehrenbach, Heide, and Davide Rodogno. *Humanitarian Photography.* New York: Cambridge University Press, 2015.

Fligstein, Neil, and Doug McAdam. "Putting Values and Institutions Back into the Theory of Strategic Action Fields: Response to Goldstone and Useem." *Sociological Theory* 30, no. 1 (2012): 48–50.

Fligstein, Neil, and Doug McAdam. "Toward a General Theory of Strategic Action Fields." *Sociological Theory* 29, no. 1 (2011): 1–26.

Fox, A. M., S. S. Jackson, N. B. Hansen, N. Gasa, M. Crewe, and K. J. Sikkema. "In Their Own Voices: A Qualitative Study of Women's Risk for Intimate Partner Violence and HIV in South Africa." *Violence Against Women* 13, no. 58 (2007): 583–602.

Fraser, Nancy. *Scales of Justice: Reimagining Political Space in a Globalizing World.* New York: Columbia University Press, 2009.

Fraser, Nancy. "Transnationalizing the Public Sphere: On the Legitimacy and Efficacy of Public Opinion in a Post-Westphalian World." *Theory, Culture & Society* 24, no. 4 (2007): 7–30.

Frickel, Scott, Sahra Gibbon, Jeff Howard, Joanna Kempner, Gwen Ottinger, and David J. Hess. "Undone Science: Charting Social Movement and Civil Society Challenges to Research Agenda Setting." *Science, Technology, & Human Values* 35, no. 4 (2010): 444–473.

Fuller, Christopher J. *See It Shoot It: The Secret History of the CIA's Lethal Drone Program.* New Haven, CT: Yale University Press, 2017.

Fuller, R. Buckminster. *Operating Manual for Spaceship Earth.* Originally published 1923. Estate of R. Buckminster Fuller, 2008.

Gabrys, Jennifer. *Program Earth: Environmental Sensing Technology and the Making of a Computational Planet.* Minneapolis: University of Minnesota Press, 2016.

Gaby, Sarah, and Neal Caren. "Occupy Online: How Cute Old Men and Malcolm X Recruited 400,000 US Users to OWS on Facebook." *Social Movement Studies* 11, no. 3–4 (2012): 367–374.

Gandhi, K. "Charles Ellet, Jr., the Pioneer American Suspension Bridge Builder." Paper presented at the *Sustainable Bridge Structures*: Proceedings of the 8th New York City Bridge Conference, New York City, August 24–25, 2015.

Gaver, William W. "Technology Affordances." In *Proceedings of the SIGCHI Conference on Human Factors in Computing Systems*, 139–146. New York: ACM, 1991.

Gibson, James J. *The Ecological Approach to Visual Perception: Classic Edition.* New York: Psychology Press, 2014.

Giddens, Anthony. *Central Problems in Social Theory.* Berkeley: University of California Press, 1979.

Giddens, Anthony. *New Rules of Sociological Method: A Positive Critique of Interpretative Sociologies.* New York: John Wiley & Sons, 2013.

Gillespie, Tarleton, Pablo J. Boczkowski, and Kirsten A. Foot. *Media Technologies: Essays on Communication, Materiality, and Society.* Cambridge, MA: MIT Press, 2014.

Gillispie, Charles Coulston. *The Montgolfier Brothers and the Invention of Aviation 1783–1784: With a Word on the Importance of Ballooning for the Science of Heat and the Art of Building Railroads.* Princeton, NJ: Princeton University Press, 2014.

Goldberg, Vicki. *The Power of Photography: How Photographs Changed Our Lives.* New York: Abbeville Press, 1991.

Goodwin, Jeff, and James M. Jasper. "Caught in a Winding, Snarling Vine: The Structural Bias of Political Process Theory." *Sociological Forum* 14, no. 1 (1999): 27–54.

Gordon, Uri. "Anarchism and the Politics of Technology." *WorkingUSA* 12, no. 3 (2009): 489–503.

Graham, Stephen. "The Urban 'Battlespace.'" *Theory, Culture & Society* 26, no. 7–8 (2009): 278–288.

Graham, Stephen. *Vertical: The City from Satellites to Bunkers.* New York: Verso Books, 2016.

Gregory, Sam. "Cameras Everywhere: Ubiquitous Video Documentation of Human Rights, New Forms of Video Advocacy, and Considerations of Safety, Security, Dignity and Consent." *Journal of Human Rights Practice* 2, no. 2 (2010): 191–207.

Gusterson, Hugh. *Drone: Remote Control Warfare.* Cambridge, MA: MIT Press, 2016.

Habermas, Jürgen. *The Theory of Communicative Action.* Boston: Beacon Press, 1984.

Hanson, Robin. *The Age of Em: Work, Love, and Life When Robots Rule the Earth.* New York: Oxford University Press, 2016.

Haraway, Donna. *Simians, Cyborgs, and Women: The Reinvention of Nature.* New York: Routledge, 2013.

Haraway, Donna J. *Staying with the Trouble: Making Kin in the Chthulucene.* Durham, NC: Duke University Press, 2016.

Hart, Clive. *Kites, an Historical Survey.* Mount Vernon, NY: Paul P. Appel, 1982.

Hawes, James. "Hated in the Nation." In *Black Mirror*, edited by Charlie Brooker, 89 minutes. United Kingdom: BBC, 2016.

Hawk, David. *The Hidden Gulag: Exposing North Korea's Prison Camps.* Washington, DC: US Committee for Human Rights in North Korea, 2003.

He, Zhou. *Mass Media and Tiananmen Square.* New York: Nova Science Publishers, 1996.

Hempel, Jessie. "Inside Facebook's Ambitious Plan to Connect the Whole World." *Wired* (2016).

Herscher, Andrew. "Surveillant Witnessing: Satellite Imagery and the Visual Politics of Human Rights." *Public Culture* 26, no. 3, 74 (2014): 469–500.

Hess, David J. *Undone Science: Social Movements, Mobilized Publics, and Industrial Transitions.* Cambridge, MA: MIT Press, 2016.

Hetland, Gabriel, and Jeff Goodwin. "The Strange Disappearance of Capitalism from Social Movement Studies." In *Marxism and Social Movements*, edited by Colin Barker, Laurence Cox, John Krinsky and Alf Gunvald Nilsen, 82–102. Leiden, NL: Brill, 2013.

Hochschild, Adam. *Bury the Chains: Prophets and Rebels in the Fight to Free an Empire's Slaves.* New York: Houghton Mifflin Harcourt, 2006.

Hoople, Gordon, Austin Choi-Fitzpatrick, and Elizabeth Reddy. "Drones for Good: Interdisciplinary Project Based Learning Between Engineering and Peace Studies." *International Journal of Engineering Education* 35, no. 2 (2019): 1378–1391.

Howard, Philip N. *The Digital Origins of Dictatorship and Democracy: Information Technology and Political Islam.* New York: Oxford University Press, 2010.

Howard, Philip N., and Muzammil M. Hussain. *Democracy's Fourth Wave? Digital Media and the Arab Spring.* New York: Oxford University Press, 2013.

Howard, Philip N., and Muzammil M. Hussain. "The Role of Digital Media." *Journal of Democracy* 22, no. 3 (2011): 35–48.

Howard, Philip N., Bence Kollanyi, and Samuel Woolley. "Bots and Automation over Twitter During the US Election." Data Memo 2016.4. Oxford, UK: Project on Computational Propaganda, 2016.

Hughes, Thomas Parke. *Human-Built World: How to Think About Technology and Culture.* Chicago: University of Chicago Press, 2004.

Hughes, Thomas Parke. *Rescuing Prometheus: Four Monumental Projects That Changed the Modern World.* New York: Pantheon Books, 1998.

Hunt, Lynn. *Inventing Human Rights: A History.* New York: W.W. Norton & Co., 2007.

Hutchby, Ian. *Conversation and Technology: From the Telephone to the Internet.* Hoboken, NJ: John Wiley & Sons, 2013.

Hutchby, Ian. "Technologies, Texts and Affordances." *Sociology* 35, no. 2 (2001): 441–456.

Jameson, Fredric, and Masao Miyoshi. *The Cultures of Globalization.* Durham, NC: Duke University Press, 1998.

Jasanoff, Sheila. *States of Knowledge: The Co-Production of Science and the Social Order.* New York: Routledge, 2004.

Jenkins, J. Craig, and Charles Perrow. "Insurgency of the Powerless: Farm Worker Movements (1946–1972)." *American Sociological Review* (1977): 249–268.

Jones, Steve. *Revolutionary Science: Transformation and Turmoil in the Age of the Guillotine.* New York: Pegasus, 2017.

Kadaba, Mrn P., H. K. Ramakrishnan, and M. E. Wootten. "Measurement of Lower Extremity Kinematics During Level Walking." *Journal of Orthopaedic Research* 8, no. 3 (1990): 383–392.

Kaiser, Philipp, and Miwon Kwon. *Ends of the Earth: Land Art to 1974*. New York: Prestel Publishing, 2012.

Kallinikos, Jannis, Paul M. Leonardi, and Bonnie A. Nardi. "The Challenge of Materiality: Origins, Scope, and Prospects." In *Materiality and Organizing: Social Interaction in a Technological World*, edited by Paul M. Leonardi, Bonnie A. Nardi, and Jannis Kallinikos, 3–22. Oxford, UK: Oxford University Press, 2012.

Kaptelinin, Victor, and Bonnie A. Nardi. *Acting with Technology: Activity Theory and Interaction Design*. Cambridge, MA: MIT Press, 2006.

Kastner, Jeffrey, and Brian Wallis. *Land and Environmental Art*. London: Phaidon Press, 1998.

Keck, Margaret E., and Kathryn Sikkink. *Activists Beyond Borders: Advocacy Networks in International Politics*. Ithaca, NY: Cornell University Press, 2014.

King, Brayden G., and Sarah A. Soule. "Social Movements as Extra-Institutional Entrepreneurs: The Effect of Protests on Stock Price Returns." *Administrative Science Quarterly* 52, no. 3 (2007): 413–442.

Kitschelt, Herbert P. "Political Opportunity Structures and Political Protest: Anti-Nuclear Movements in Four Democracies." *British Journal of Political Science* 16, no. 01 (1986): 57–85.

Klauser, Francisco, and Silvana Pedrozo. "Power and Space in the Drone Age: A Literature Review and Politico-Geographical Research Agenda." *Geographica Helvetica* 70, no. 4 (2015): 285.

Kofman, Sarah. *Camera Obscura: Of Ideology*. Ithaca, NY: Cornell University Press, 1999.

Kondratieff, Nikolai D. "The Long Waves in Economic Life." *Review: A Journal of the Fernand Braudel Center* 2 (1979): 519–562.

Kurgan, Laura. *Close Up at a Distance: Mapping, Technology, and Politics*. Cambridge, MA: MIT Press, 2013.

Kuznets, Simon. "Economic Growth and Income Inequality." *The American Economic Review* 45, no. 1 (1955): 1–28.

Kuznets, Simon Smith. "Secular Movements in Production and Prices." Boston: Houghton Mifflin, 1930.

Lambeth, Benjamin S. *NATO's Air War for Kosovo: A Strategic and Operational Assessment*. Santa Monica, CA: Rand Corporation, 2001.

Lanier, Jaron. *Who Owns the Future?* New York: Simon & Schuster, 2014.

Latour, Bruno. *Reassembling the Social: An Introduction to Actor-Network-Theory.* New York: Oxford University Press, 2005.

Leach, Darcy K. "The Iron Law of What Again? Conceptualizing Oligarchy across Organizational Forms." *Sociological Theory* 43 (2005): 312–337.

Lefebvre, Henri, and Donald Nicholson-Smith. *The Production of Space.* New York: Oxford Blackwell, 1991.

Lehmann, Ernst August, Leonhard Adelt, Jay Dratler, and Charles Emery Rosendahl. *Zeppelin: The Story of Lighter-Than-Air Craft.* London: Longmans, Green and Company, 1937.

Leonardi, Paul M., Bonnie A. Nardi, and Jannis Kallinikos. *Materiality and Organizing: Social Interaction in a Technological World.* Oxford, UK: Oxford University Press, 2012.

Lin, Nan. *The Struggle for Tiananmen: Anatomy of the 1989 Mass Movement.* Westport, CT: Praeger Publishers, 1992.

Linfield, Susie. *The Cruel Radiance: Photography and Political Violence.* Chicago: University of Chicago Press, 2011.

Little, Andrew T. "Communication Technology and Protest." *The Journal of Politics* 78, no. 1 (2016): 152–166.

Livingston, Steven. "Digital Affordances and Human Rights Advocacy." Berlin: DFG Collaborative Research Center, 2016.

Livingston, Steven, and Douglas A. Van Belle. "The Effects of Satellite Technology on Newsgathering from Remote Locations." *Political Communication* 22, no. 1 (2005): 45–62.

Livingston, Steven, and Gregor Walter-Drop. *Bits and Atoms: Information and Communication Technology in Areas of Limited Statehood.* New York: Oxford University Press, 2013.

Luders, Joseph E. *The Civil Rights Movement and the Logic of Social Change.* New York: Cambridge University Press, 2010.

Lyon, David. *The Electronic Eye: The Rise of Surveillance Society.* Minneapolis: University of Minnesota Press, 1994.

MacKenzie, Donald A. *Knowing Machines: Essays on Technical Change.* Cambridge, MA: MIT Press, 1998.

Mansell, Claudia Martinez. "A Change of Perspective: Aerial Photography and 'the Right to the City' in a Palestinian Refugee Camp." In *Visual Imagery and Human Rights Practice*, 213–228. New York: Palgrave Macmillan, 2018.

Marres, Noortje, and Javier Lezaun. "Materials and Devices of the Public: An Introduction." *Economy and Society* 40, no. 4 (2011): 489–509.

Marvin, Carolyn. *When Old Technologies Were New: Thinking About Electric Communication in the Late Nineteenth Century.* New York: Oxford University Press, 1990.

Marx, Karl. *The 18th Brumaire of Louis Bonaparte.* Cabin John, MD: Wildside Press, 2008.

McAdam, Doug. *Political Process and the Development of Black Insurgency, 1930–1970.* Chicago: University of Chicago Press, 2010.

McCarthy, John D., and Mayer N. Zald. "Resource Mobilization and Social Movements: A Partial Theory." *American Journal of Sociology* 82, no. 6 (1977): 1212–1241.

McDonnell, Terence E. *Best Laid Plans: Cultural Entropy and the Unraveling of Aids Media Campaigns.* Chicago: University of Chicago Press, 2016.

McDonnell, Terence E. "Cultural Objects as Objects: Materiality, Urban Space, and the Interpretation of Aids Campaigns in Accra, Ghana." *American Journal of Sociology* 115, no. 6 (2010): 18–52.

McPherson, Ella. "Digital Human Rights Reporting by Civilian Witnesses: Surmounting the Verification Barrier." In *Produsing Theory in a Digital World 2.0: The Intersection of Audiences and Production in Contemporary Theory*, edited by Rebecca Ann Lind, 193–209. New York: Peter Lang, 2015.

McVeigh, Rory. *The Rise of the Ku Klux Klan: Right-Wing Movements and National Politics.* Minneapolis: University of Minnesota Press, 2009.

Meier, Patrick. *Digital Humanitarians: How Big Data Is Changing the Face of Humanitarian Response.* Boca Raton, FL: CRC Press, 2015.

Memou, Antigoni. *Photography and Social Movements: From the Globalisation of the Movement (1968) to the Movement Against Globalisation (2001).* Manchester, UK: Manchester University Press, 2015.

Merton, Robert King, and Robert C Merton. *Social Theory and Social Structure.* New York: Simon and Schuster, 1968.

Michel, Arthur Holland. "Counter-Drone Systems." Annandale-on-Hudson, NY: Center for the Study of the Drone at Bard College, 2018.

Michel, Arthur Holland, and Dan Gettinger. "Drone Year in Review." Annandale-on-Hudson, NY: Center for the Study of the Drone at Bard College, 2017.

Michels, Robert. *Political Parties: A Sociological Study of the Oligarchical Tendencies of Modern Democracy.* New York: Hearst's International Library Company, 1915.

Milan, Stefania. *Social Movements and Their Technologies: Wiring Social Change.* New York: Palgrave Macmillan, 2013.

Mills-Scofield, Deborah. "It's Not Just Semantics: Managing Outcomes vs. Outputs." *Harvard Business Review,* November 26, 2012.

Morozov, Evgeny. *The Net Delusion: The Dark Side of Internet Freedom.* New York: Public Affairs, 2011.

Morozov, Evgeny. *To Save Everything, Click Here: The Folly of Technological Solutionism.* New York: PublicAffairs, 2013.

Morris, Robert. "Aligned with Nazca." *Artforum* 14, no. 2 (1975): 26–39.

Murphy, Robin R. *Disaster Robotics.* Cambridge, MA: MIT Press, 2014.

Nadelmann, Ethan A. "Global Prohibition Regimes: The Evolution of Norms in International Society." *International Organization* 44, no. 04 (1990): 479–526.

Nagy, Peter, and Gina Neff. "Imagined Affordance: Reconstructing a Keyword for Communication Theory." *Social Media + Society* 1, no. 2 (2015).

Needham, Joseph. *Science and Civilisation in China, Vol 4. Physics and Physical Technology. Part II: Mechanical Engineering.* London: Cambridge University Press, 1965.

Neff, Gina, Tim Jordan, Joshua McVeigh-Schultz, and Tarleton Gillespie. "Affordances, Technical Agency, and the Politics of Technologies of Cultural Production." *Journal of Broadcasting and Electronic Media* 56, no. 2 (2012): 299–313.

Neff, Gina, and Dawn Nafus. *The Quantified Self.* Cambridge, MA: MIT Press, 2016.

Newhall, Beaumont. *Airborne Camera; the World from the Air and Outer Space.* New York: Hastings House, 1969.

Newman, Lee, and Jay Hartley Newman. *Kite Craft: The History and Processes of Kitemaking Throughout the World.* New York: Crown, 1988.

Norman, Donald A. *The Design of Everyday Things: Revised and Expanded Edition.* New York: Basic Books, 2013.

Oldfield, John R. *Popular Politics and British Anti-Slavery: The Mobilisation of Public Opinion Against the Slave Trade 1787–1807.* New York: Routledge, 2012.

Olson, Mancur. *The Logic of Collective Action: Public Goods and the Theory of Groups.* Cambridge, MA: Harvard University Press, 2009.

O'Neil, Cathy. *Weapons of Math Destruction: How Big Data Increases Inequality and Threatens Democracy.* New York: Broadway Books, 2017.

Outwater, A., N. Abrahams, and J. C. Campbell. "Women in South Africa: Intentional Violence and HIV/AIDS: Intersections and Prevention." *Journal of Black Studies* 35, no. 135 (2005): 135–154.

Owen, Taylor. *Disruptive Power: The Crisis of the State in the Digital Age.* New York: Oxford University Press, 2015.

Papacharissi, Zizi. *Affective Publics: Sentiment, Technology, and Politics.* New York: Oxford University Press, 2015.

Parks, Lisa. *Cultures in Orbit: Satellites and the Televisual.* Durham, NC: Duke University Press, 2005.

Parks, Lisa. "Digging into Google Earth: An Analysis of 'Crisis in Darfur.'" *Geoforum* 40, no. 4 (2009): 535–545.

Parks, Lisa. "Satellite Views of Srebrenica: Tele-Visuality and the Politics of Witnessing." *Social Identities* 7, no. 4 (2001): 585–611.

Parks, Lisa, and Caren Kaplan. *Life in the Age of Drone Warfare.* Durham, NC: Duke University Press, 2017.

Parks, Lisa, and Nicole Starosielski. *Signal Traffic: Critical Studies of Media Infrastructures.* Urbana, IL: University of Illinois Press, 2015.

Peters, Benjamin. *How Not to Network a Nation: The Uneasy History of the Soviet Internet.* Cambridge, MA: MIT Press, 2016.

Pierskalla, Jan H., and Florian M. Hollenbach. "Technology and Collective Action: The Effect of Cell Phone Coverage on Political Violence in Africa." *American Political Science Review* 107, no. 2 (2013): 207–224.

Pike, Sid. *We Changed the World: Memoirs of a CNN Global Satellite Pioneer.* St. Paul, MN: Paragon House, 2005.

Piketty, Thomas. "About Capital in the Twenty-First Century." *American Economic Review* 105, no. 5 (2015): 48–53.

Piven, Frances Fox, and Richard A. Cloward. *Poor People's Movements: Why They Succeed, How They Fail.* New York: Vintage Books, 1979.

Pocock, George. *The Aeropleustic Art or Navigation in the Air, by the Use of Kites or Buoyant Sails.* London: Sherwood & Company, 1827.

Polletta, Francesca, and Kelsy Kretschmer. "'Free Spaces' in Collective Action." *Theory and Society* 28, no. 1 (1999): 1–38.

Qiang, Xiao. "The Battle for the Chinese Internet." *Journal of Democracy* 22, no. 2 (2011): 47–61.

Rawls, John. *A Theory of Justice.* Cambridge, MA: Harvard University Press, 2009.

Reddy, Elizabeth, Gordon Hoople, and Austin Choi-Fitzpatrick. "Boundary Objects in Classroom Practice: Using Drones to Foster Critical Engagement." *Journal of Engineering Studies* 11, no. 1 (2019): 51–64.

Reynolds, Craig W. "Flocks, Herds and Schools: A Distributed Behavioral Model." Paper presented at the ACM SIGGRAPH Conference on Computer Graphics and Interactive Techniques. New York: ACM, 1987.

Riggle, Nick. *On Being Awesome: A Unified Theory of How Not to Suck.* New York: Penguin, 2017.

Rinberg, Toly, Maya Anjur-Dietrich, Marcy Beck, Andrew Bergman, Justin Derry, Lindsey Dillon, Gretchen Gehrke, et al. "Changing the Digital Climate: How Climate Change Web Content Is Being Censored Under the Trump Administration." Environmental Data and Governance Initiative, 2018.

Risse, Thomas, Stephen C. Ropp, and Kathryn Sikkink. *The Persistent Power of Human Rights: From Commitment to Compliance.* New York: Cambridge University Press, 2013.

Risse, Thomas, Stephen C. Ropp, and Kathryn Sikkink. *The Power of Human Rights: International Norms and Domestic Change.* New York: Cambridge University Press, 1999.

Ristovska, Sandra. "The Rise of Eyewitness Video and Its Implications for Human Rights: Conceptual and Methodological Approaches." *Journal of Human Rights* 15, no. 3 (2016): 347–360.

Ristovska, Sandra. "Witnessing and the Failure of Communication." *The Communication Review* 17, no. 2 (2014): 143–158.

Robinson, Douglas Hill. *Giants in the Sky: A History of the Rigid Airship*. Seattle, WA: University of Washington Press, 1973.

Rochon, Thomas R. *Culture Moves: Ideas, Activism, and Changing Values*. Princeton, NJ: Princeton University Press, 2000.

Rogers, Everett M. *Diffusion of Innovations*. New York: Simon and Schuster, 2010.

Roscigno, Vincent J., and William F. Danaher. *The Voice of Southern Labor*. Minneapolis: University of Minnesota Press, 2004.

Rose, Gillian. *Visual Methodologies: An Introduction to Researching with Visual Materials*. Thousand Oaks, CA: Sage, 2016.

Sachs, Wolfgang. *Planet Dialectics: Explorations in Environment and Development*. London: Zed Books, 2015.

Sang-su, Choe. "The Survey of Korean Kites." Seoul: Korea Books Pub. Co, 1958.

Schiffer, Michael Brian. *Draw the Lightning Down: Benjamin Franklin and Electrical Technology in the Age of Enlightenment*. Berkeley: University of California Press, 2006.

Schmidt, Henrike. "From a Bird's Eye Perspective: Aerial Drone Photography and Political Protest. A Case Study of the Bulgarian #Resign Movement 2013." *Digital Icons: Studies in Russian, Eurasian and Central European New Media* 13 (2015): 1–27.

Schurman, Rachel, and William A. Munro. *Fighting for the Future of Food: Activists versus Agribusiness in the Struggle over Biotechnology*. Minneapolis: University of Minnesota Press, 2013.

Schwartz, Michael. *Radical Protest and Social Structure: The Southern Farmers' Alliance and Cotton Tenancy, 1880–1890*. Chicago: University of Chicago Press, 1988.

Scott, James C. *The Art of Not Being Governed: An Anarchist History of Upland Southeast Asia*. New Haven, CT: Yale University Press, 2014.

Scott, James C. *Seeing Like a State: How Certain Schemes to Improve the Human Condition Have Failed*. New Haven, CT: Yale University Press, 1998.

Scott, James C. *Two Cheers for Anarchism*. Princeton, NJ: Princeton University Press, 2012.

Scott, James C. *Weapons of the Weak: Everyday Forms of Peasant Resistance*. New Haven, CT: Yale University Press, 2008.

Sealy, Mark, Roger Malbert, and Alice Lobb. *Documenting Disposable People: Contemporary Global Slavery.* London: Hayward Gallery, 2008.

Segal, Rafi, David Tartakover, and Eyal Weizman. *A Civilian Occupation: The Politics of Israeli Architecture.* New York: Verso, 2003.

Sharp, Gene. *The Politics of Nonviolent Action.* 3 Vols. Boston: Porter Sargent (1973).

Shaw, Ian GR. *Predator Empire: Drone Warfare and Full Spectrum Dominance.* Minneapolis: University of Minnesota Press, 2016.

Silverman, Kaja. *The Miracle of Analogy, or, the History of Photography.* Stanford, CA: Stanford University Press, 2015.

Singer, Peter Warren. *Wired for War: The Robotics Revolution and Conflict in the 21st Century.* New York: Penguin, 2009.

Sliwinski, Sharon. "The Childhood of Human Rights: The Kodak on the Congo." *Journal of Visual Culture* 5, no. 3 (2006): 333–363.

Sliwinski, Sharon. *Human Rights in Camera.* Chicago: University of Chicago Press, 2011.

Sloterdijk, Peter. "Airquakes." *Environment and Planning D: Society and Space* 27, no. 1 (2009): 41–57.

Sloterdijk, Peter. *Terror from the Air.* Los Angeles: Semiotext(e), 2009.

Sloterdijk, Peter. *Versprechen Auf Deutsch.* Vol. 631. Frankfurt: Suhrkamp, 1990.

Smihula, Daniel. "The Waves of the Technological Innovations of the Modern Age and the Present Crisis as the End of the Wave of the Informational Technological Revolution." *Studia Politica Slovaca*, no. 1 (2009): 32–47.

Smithson, Robert. *Robert Smithson: The Collected Writings.* Berkeley: University of California Press, 1996.

Smithson, Robert. "Towards the Development of an Air Terminal Site." *Artforum* 6, no. 10 (1967): 52–60.

Snow, David A. "Social Movements as Challenges to Authority: Resistance to an Emerging Conceptual Hegemony." In *Authority in Contention*, edited by Daniel Cress and Daniel J Myers, 3–25. Bingley, UK: Emerald Group Publishing, 2004.

Sorokowski, P., A. Sorokowska, A. Oleszkiewicz, T. Frackowiak, A. Huk, and K. Pisanski. "Selfie Posting Behaviors Are Associated with Narcissism Among Men." *Personality and Individual Differences* 85 (2015): 123–127.

Soule, Sarah A. "The Student Divestment Movement in the United States and Tactical Diffusion: The Shantytown Protest." *Social Forces* 75, no. 3 (1997): 855–882.

Sprigg, Christopher. *The Airship—Its Design, History, Operation and Future.* London: Sampson Low, Marsden & Company, 1931.

Staggenborg, Suzanne. *Social Movements.* New York: Oxford University Press, 2015.

Staples, William G. *Everyday Surveillance: Vigilance and Visibility in Postmodern Life.* 2nd ed. Lanham, MD: Rowman & Littlefield, 2014.

Starkman, Dean. *The Watchdog That Didn't Bark: The Financial Crisis and the Disappearance of Investigative Reporting.* New York: Columbia University Press, 2014.

Starosielski, Nicole. *The Undersea Network.* Durham, NC: Duke University Press, 2015.

Suay, Juan Miguel, and David Teira. "Kites." *Nuncius* 29, no. 2 (2014): 439–463.

Swidler, Ann. "Culture in Action: Symbols and Strategies." *American Sociological Review* (1986): 273–286.

Tarrow, Sidney G. *Power in Movement: Social Movements and Contentious Politics.* New York: Cambridge University Press, 2011.

Taylor, Michael John Haddrick. *Brassey's World Aircraft & Systems Directory, 1999/2000.* London: Brassey's, 1999.

Taylor, Verta, and Nella Van Dyke. "'Get up, Stand Up': Tactical Repertoires of Social Movements." In *The Blackwell Companion to Social Movements*, edited by Hanspeter Kriesi, Sarah A. Soule, and David A. Snow, 262–293. Malden, MA: Blackwell, 2004.

Tilly, Charles. "From Interactions to Outcomes in Social Movements." In *How Social Movements Matter,* edited by Marco Giugni, Doug McAdam, and Charles Tilly, 253–270. Minneapolis: University of Minnesota Press, 1999.

Tilly, Charles. *From Mobilization to Revolution.* New York: McGraw-Hill, 1978.

Tilly, Charles. *Regimes and Repertoires.* Chicago: University of Chicago Press, 2010.

Tilly, Charles. "Spaces of Contention." *Mobilization* 5, no. 2 (2000): 135–159.

Tilly, Charles, and Lesley J. Wood. *Social Movements 1768–2012.* New York: Routledge, 2015.

Traugott, Mark. *The Insurgent Barricade.* Berkeley: University of California Press, 2010.

Tufekci, Zeynep. *Twitter and Tear Gas: The Power and Fragility of Networked Protest.* New Haven, CT: Yale University Press, 2017.

Unger, Roberto Mangabeira. *The Self Awakened: Pragmatism Unbound.* Cambridge, MA: Harvard University Press, 2007.

Vaidyanathan, Brandon, Michael Strand, Austin Choi-Fitzpatrick, Thomas Buschman, Meghan Davis, and Amanda Varela. "Causality in Contemporary American Sociology: An Empirical Assessment and Critique." *Journal for the Theory of Social Behaviour* 46, no. 1 (2016): 3–26.

Ventry, Baron Arthur Frederick Daubeney Eveleigh-de Moleyns and Eugène M. Koleśnik. *Airship Saga: The History of Airships Seen through the Eyes of the Men Who Designed, Built, and Flew Them.* Poole, Dorset, UK: Blandford Press, 1982.

Vera, Lourdes A., Lindsey Dillon, Sara Wylie, Jennifer Liss Ohayon, Aaron Lemelin, Phil Brown, Christopher Sellers, et al. "Data Resistance: A Social Movement Organizational Autoethnography of the Environmental Data and Governance Initiative." *Mobilization:* 23, no. 4 (2018): 511–529.

Virilio, Paul. *The Information Bomb.* Trans. Chris Turner. New York: Verso, 2000.

Virilio, Paul. *The Vision Machine.* Bloomington: Indiana University Press, 1994.

Virilio, Paul. *War and Cinema: The Logistics of Perception.* New York: Verso, 1989.

Virilio, Paul, and Benjamin H. Bratton. *Speed and Politics.* Los Angeles: Semiotext(e), 2006.

Virilio, Paul, and Philippe Petit. *Politics of the Very Worst.* Trans. Sylvère Lotringer. New York: Semiotext(e), 1999.

Vygotsky, Lev Semenovich. *Mind in Society: The Development of Higher Psychological Processes.* Cambridge, MA: Harvard University Press, 1980.

Walder, Andrew G. "Political Sociology and Social Movements." *Annual Review of Sociology* 35 (2009): 393-412.

Wallerstein, Immanuel. *The Modern World-System I: Capitalist Agriculture and the Origins of the European World-Economy in the Sixteenth Century.* Berkeley: University of California Press, 1974.

Wang, Dan J., and Sarah A. Soule. "Social Movement Organizational Collaboration: Networks of Learning and the Diffusion of Protest Tactics, 1960–1995." *American Journal of Sociology* 117, no. 6 (2012): 1674–1722.

Warren, Jeffrey Yoo. "Grassroots Mapping: Tools for Participatory and Activist Cartography." Thesis. Massachusetts Institute of Technology, 2010.

Weightman, Gavin. *Signor Marconi's Magic Box: The Invention That Sparked the Radio Revolution*. New York: HarperCollins, 2003.

Weizman, Eyal. *Hollow Land: Israel's Architecture of Occupation*. New York: Verso Books, 2012.

Wertsch, James V. *Mind as Action*. New York: Oxford University Press, 1998.

Whittle, Richard. *Predator: The Secret Origins of the Drone Revolution*. New York: Macmillan, 2014.

Winner, Langdon. *Autonomous Technology: Technics-out-of-Control as a Theme in Political Thought*. MIT Press, 1978.

Winner, Langdon. "Do Artifacts Have Politics?" *Daedalus* (1980): 121–136.

Winner, Langdon. *The Whale and the Reactor: A Search for Limits in an Age of High Technology*. Chicago: University of Chicago Press, 2010.

Wollheim, Peter. "Towards a Critical History of the 35mm Still Photographic Camera in North America, 1896 to 1980." Thesis. University of California, Los Angeles, 1993.

Woodman, Jim. *Nazca: Journey to the Sun*. New York: Pocket, 1977.

Woods, Chris. *Sudden Justice: America's Secret Drone Wars*. New York: Oxford University Press, 2015.

Yamada, Takayuki, Seiichi Gohshi, and Isao Echizen. "Privacy Visor: Method Based on Light Absorbing and Reflecting Properties for Preventing Face Image Detection." In *IEEE International Conference on Systems, Man, and Cybernetics*, 1572–1577. Washington, DC: IEEE Computer Society, 2013.

Young, Michael P. *Bearing Witness Against Sin: The Evangelical Birth of the American Social Movement*. Chicago: University of Chicago Press, 2006.

Zeller, Bob. *The Blue and Gray in Black and White: A History of Civil War Photography*. Westport, CT: Greenwood Publishing Group, 2005.

INDEX

Printed in the United States
by Baker & Taylor Publisher Services

Printed in the United States
by Baker & Taylor Publisher Services